C0-AYJ-927

Springer Series in Information Sciences 19

Editor: Manfred R. Schroeder

Springer Series in Information Sciences

1 **Content-Addressable Memories**
By T. Kohonen 2nd Edition

2 **Fast Fourier Transform and Convolution Algorithms**
By H. J. Nussbaumer 2nd Edition

3 **Pitch Determination of Speech Signals**
Algorithms and Devices By W. Hess

4 **Pattern Analysis** By H. Niemann
2nd Edition

5 **Image Sequence Analysis**
Editor: T.S. Huang

6 **Picture Engineering**
Editors: King-sun Fu and T.L. Kunii

7 **Number Theory in Science and Communication** With Applications in Cryptography, Physics, Digital Information, Computing, and Self-Similarity By M.R. Schroeder
2nd Edition

8 **Self-Organization and Associative Memory** By T. Kohonen 3rd Edition

9 **Digital Picture Processing**
An Introduction By L.P. Yaroslavsky

10 **Probability, Statistical Optics and Data Testing** A Problem Solving Approach By B.R. Frieden

11 **Physical and Biological Processing of Images** Editors: O.J. Braddick and A.C. Sleigh

12 **Multiresolution Image Processing and Analysis** Editor: A. Rosenfeld

13 **VLSI for Pattern Recognition and Image Processing** Editor: King-sun Fu

14 **Mathematics of Kalman-Bucy Filtering**
By P.A. Ruymgaart and T.T. Soong
2nd Edition

15 **Fundamentals of Electronic Imaging Systems** Some Aspects of Image Processing By W.F. Schreiber

16 **Radon and Projection Transform-Based Computer Vision**
Algorithms, A Pipeline Architecture, and Industrial Applications By J.L.C. Sanz, E.B. Hinkle, and A.K. Jain

17 **Kalman Filtering** with Real-Time Applications By C.K. Chui and G. Chen

18 **Linear Systems and Optimal Control**
By C.K. Chui and G. Chen

19 **Harmony: A Psychoacoustical Approach** By R. Parncutt

20 **Group Theoretical Methods in Image Understanding** By Ken-ichi Kanatani

21 **Linear Prediction Theory**
A Mathematical Basis for Adaptive Systems By P. Strobach

Richard Parncutt

Harmony: A Psychoacoustical Approach

With 22 Figures

Springer-Verlag Berlin Heidelberg New York
London Paris Tokyo Hong Kong

Dr. Richard Parncutt

Departments of Music and Psychology,
University of New England, Armidale NSW 2351, Australia

Series Editors:
Professor Thomas S. Huang

Department of Electrical Engineering and Coordinated Science Laboratory,
University of Illinois, Urbana, IL 61801, USA

Professor Teuvo Kohonen

Department of Technical Physics, Helsinki University of Technology,
SF-02150 Espoo 15, Finland

Professor Dr. Manfred R. Schroeder

Drittes Physikalisches Institut, Universität Göttingen, Bürgerstrasse 42–44,
D-3400 Göttingen, Fed. Rep. of Germany

Managing Editor: Helmut K.V. Lotsch

Springer-Verlag, Tiergartenstrasse 17,
D-6900 Heidelberg, Fed. Rep. of Germany

ISBN 3-540-51279-9 Springer-Verlag Berlin Heidelberg New York
ISBN 0-387-51279-9 Springer-Verlag New York Berlin Heidelberg

Library of Congress Cataloging-in-Publication Data. Parncutt, Richard. 1957–. Harmony: a psychoacoustical approach / Richard Parncutt. p. cm. – (Springer series in information sciences: 19) Bibliography: p. Includes index. 1. Music – Psychology. 2. Harmony. 3. Music – Acoustics and physics. I. Title. II. Series. ML3836.P3 1989 781.2'5'019–dc19 89-11555

Printed in the United States of America

Typesetting: K & V Fotosatz, 6124 Beerfelden
2154/3150-543210 – Printed on acid-free paper

To the pacifists

Preface

My first encounter with the theory of harmony was during my last year at school (1975). This fascinating system of rules crystallized the intuitive knowledge of harmony I had acquired from years of piano playing, and facilitated memorization, transcription, arrangement and composition. For the next five years, I studied music (piano) and science (physics) at the University of Melbourne. This "strange combination" started me wondering about the *origins* of those music theory "rules". To what extent were they determined or influenced by physics? mathematics? physiology? conditioning?

In 1981, the supervisor of my honours project in musical acoustics, Neville Fletcher, showed me an article entitled "Pitch, consonance, and harmony", by a certain Ernst Terhardt of the Technical University of Munich. By that stage, I had devoured a considerable amount of (largely unsatisfactory) material on the nature and origins of harmony, which enabled me to recognize the significance of Terhardt's article. But it was not until I arrived in Munich the following year (on Terhardt's invitation) that I began to appreciate the consequences of his "psychoacoustical" approach for the theory of harmony. That is what this book is about.

The book presents Terhardt's work against the broad context of music perception research, past and present. Music perception is a multidisciplinary mixture of physics, psychology and music. Where different theoretical approaches appear contradictory, I try to show instead that they complement and enrich one another. Readers are assumed to be acquainted with basic principles of harmony, acoustics (including spectral analysis), and computer modelling.

The book is based on my Ph.D. thesis, which was submitted in 1986 at the University of New England, Armidale NSW, Australia. I am indebted to my supervisors, Neville Fletcher (Physics), Catherine Ellis (Music) and William Noble (Psychology), for their active interest, informed criticism and general guidance; to Gerhard Stoll and David Heap, for helping to set up the experiments; to my Ph.D. examiners, Howard Pollard (Physics, University of New South Wales), Alan Costall (Psychology, University of Southampton) and Jeff Pressing (Music, La Trobe University), for helpful suggestions, criticism and encouragement; and to Neil Buckland, for music-oriented comments.

The theoretical content of the book was influenced and inspired by conversations with staff of the Institute of Electroacoustics, TU Munich, and with academics and students met at conferences and visited during and after my Ph.D. Special thanks to Juan Roederer for organizing the Workshops on the

Physics and Neuropsychology of Music at Ossiach, Austria, which I attended in 1983 and 1985.

Most of all I would like to thank Ernst Terhardt, for his hospitality, approachability and critical guidance during my stay in Munich in 1982–1983. I hope with this book to bring his work to the attention of a wider public, and thereby to contribute to the demystification of the harmonic conventions of Western music.

Armidale, May 1988 *Richard Parncutt*

Contents

1. Background

The conventions of mainstream Western harmony may be divided into conventions for writing individual chords (stacked thirds on a root; inversion, doubling, spacing), pairs of chords (harmonic relationship, voice leading, avoidance of parallels), and chord progressions (scales, keys, tonality). A range of scientific and pseudoscientific theories have been advanced to explain harmonic conventions, based on frequency ratios, the harmonic series, beats, combination tones, periodicity, cognitive structures, generative grammars, and mathematical groups. Terhardt [1974a, 1976] explained the perception of harmony in terms of perceptual familiarity with the pitch pattern of typical complex tones. Unlike other theories, Terhardt's is both empirically based and capable of explaining harmony's origins.

1.1 Music Theory

1.1.1 Introduction

Music theory comprises conventions ("rules") and guidelines for arranging, composing and analyzing music. The music theory of interest in this study is that of diatonic Western music: music in major and minor keys from the Baroque, Classical and Romantic periods. Of particular interest is the theory of chord progressions in three or more voices, i.e., with three or more notes per chord.

All rules can be broken. Perhaps the most successful composers were those who broke the so-called rules of music theory most effectively. In order to maintain stylistic unity, however, it is necessary most of the time to adhere to appropriate compositional conventions.

The conventions of music theory are generally formulated by theorists long after composers have begun to use them consistently. For example, the "rules" of harmony and form in eighteenth century music were largely developed in the nineteenth century. Successful composers in the eighteenth century tended to compose by intuition and direct imitation. In the nineteenth century, music-theoretical formulas were developed by which budding composers could imitate the harmonic styles of their predecessors in a more general and organized fashion. The time lag between practice and theory, and the tendency for the theory to be created by theorists (rather than "great" composers), make music theory a somewhat idealized and simplified form of musical reality.

The conventions of diatonic music are explained in a variety of texts [e.g. Schoenberg 1911; MacPherson 1920; Hindemith 1940; Forte 1962; Goldman 1965; Aldwell and Schachter 1978; Piston 1978]. The most important of these conventions are introduced and summarized, along with brief accounts of their historical origins, in the following sections.

1.1.2 Single Chords

A chord is a simultaneity of three or more notes. In Western music theory, chords are constructed by superposing thirds on a note called the root (Fig. 1.1 a). Superposing two thirds produces a triad of root, third and fifth. The major and minor triads are regarded as the most consonant triads, because their fifths are "perfect" (7 semitones). Other possible constructions are the diminished and augmented triads, named after the quality of their fifths, and the suspended triad, named after its suspended fourth (which normally resolves to a third). A seventh chord is formed by adding a third above the fifth

Fig. 1.1. Basic structures in music theory

of a triad. The most consonant seventh chords in Western music theory are those with perfect fifths above the root: the major-minor (or dominant) seventh chord (with a major third and a minor seventh above the root), the minor seventh chord (minor third, minor seventh) and the major seventh chord (major third, major seventh). Seventh chords with diminished fifths include the half-diminished seventh (minor third, diminished fifth, minor seventh) and the diminished seventh (minor third, diminished fifth, diminished seventh).

Music theory assumes *octave equivalence* (or octave generalization): notes one or more octaves apart are assumed to have the same harmonic function. Such notes are said (in music theory) to belong to the same *pitch class* [Babbitt 1955], or (in music psychology) to have the same *chroma.* Similarly, a *chord class* is specified by the pitch classes of its notes. Different *voicings* of a chord class have different bass notes (inversion), spacing and doubling.

Chords are *inverted* by transposing one of the upper notes down an octave into the bass (i.e. lowest) part. Major and minor triads most commonly appear in root position (i.e. not inverted); and first inversions (third in the bass) occur more commonly than second inversions (fifth in the bass). Diminished triads are most often heard in first inversion. Augmented triads (and also diminished seventh chords) are enharmonically equivalent in all inversions, and – partly because of this ambiguity – are relatively rare in diatonic music. Seventh chords appear in all inversions, but some inversions, like the root position of the major-minor seventh, minor seventh and major seventh chords, the first inversion of the minor seventh (producing a major added sixth), and the first inversion of the half-diminished seventh (producing a minor added sixth), are more usual.

The commonly taught conventions of *spacing* of chords are as follows. The intervals between neighbouring upper notes (in the treble staff) should not exceed one octave. Intervals between lower notes (in the bass staff) may exceed an octave, and should not be too small: thirds in the lower bass region are quite dissonant.

Four-part music largely comprises triads in which one note has been *doubled* at the octave, producing two notes with the same chroma in the one chord. It is generally recommended that the root of a major or minor triad be doubled, and that the third of a major triad, or the lowest note (the conventional root) of a diminished triad, should not. In any case, the leading note (scale degree VII, see Fig. 1.1 b) is not normally doubled in four-part writing. The most commonly doubled notes fall on tonally strong scale degrees such as I, V, IV and II (in order of decreasing importance). For example, the best note to double in an "augmented sixth" chord on ♭VI[1] is not the conventional root of the chord (♭VI) but the major third (I), because I is so much more stable than ♭VI (and also because parallels are avoided this way, see below).

Rules for writing single chords were in evidence, at least intuitively, in the Middle Ages. The number of pitch classes normally used in chords in homo-

[1] Flat sixth scale degree (flat submediant)

phonic music increased from one (octave-unison) through two (open fifths or fourths, as in some *organum* styles) to three (major and minor triads) during these centuries [Randel 1971]. Conventions of doubling and spacing also developed during this time. Chords of four pitch classes (tetrads, or sevenths) appeared in passing in contrapuntal Medieval and Renaissance styles, but were not recognized as harmonies (albeit as dissonant ones) until the early Baroque in the works of Monteverdi [Arnold and Fortune 1968].

1.1.3 Pairs of Chords

The contrapuntal techniques of the Middle Ages and Renaissance provided the basis from which conventions for writing sequential pairs of chords in the Renaissance and Baroque periods developed. For example, the emergence of what we now know as triads and perfect (V-I) cadences in the fifteenth century may be traced to medieval conventions for approaching perfect consonances, e.g. contrary motion from the major sixth to the octave; a shift from three- to four-part harmony; and increasingly strict prohibition of parallel fifths and octaves [Randel 1971]. These developments eventually led to the establishment of the following music-theoretical concepts.

Sequential chords sound related, or "progress" well, if they stand in a strong *harmonic relationship* with each other. In music theory, the nature of harmonic relationship is explained indirectly, in terms of the diatonic scales and the cycle of fifths (Fig. 1.1 b, c). The *diatonic scales* include major, harmonic minor and melodic minor forms. The *cycle of fifths* is a theoretical construction of chroma in which neighbours are a fifth (or fourth) apart. Chords are generally supposed to be harmonically related if all their notes belong to the same diatonic scale or if their roots are close together on the cycle of fifths.

Sequential chords generally progress well if they have notes in common or if their voices move by step (major or minor seconds) rather than by leap (thirds, fourths, etc.). This aspect of the relationship between sequential chords is called *voice leading*, or *melodic relationship*. Leaps which do occur preferably cover harmonically simple intervals such as the octave, perfect fourth or perfect fifth. The disruptive effect of a large leap is reduced if the note following the leap falls between the pitches of the last two notes – preferably a step away from the second note.

When two voices proceed from one chord to the next in parallel motion at a constant pitch distance, *parallels* (or *consecutives*) result. Parallel fifths and octaves were common in the music of the Middle Ages (e.g. *organum*) but were generally avoided in the three- and four-part writing of the Baroque, Classical and Romantic periods.

It is useful to think of *tonal relationships* as combinations of harmonic and melodic relationships (or harmonic and scalar relationships [Sloboda 1985, p. 22]). For example, a chromatic chord (e.g. the diminished seventh B–D–F–A♭) may not stand in a particularly strong *harmonic* relationship with its chord of resolution (e.g. C–E–G), but if the *melodic* relationship between the chords is strong (e.g. due to the semitone relationships B–C and A♭–G) then their overall *tonal* relationship is strong.

1.1.4 Chord Progressions

The notes of chord progressions written in the Middle Ages and Renaissance largely conformed to *standard heptatonic scales* (corresponding to the white notes of the piano, or transpositions thereof). Depending on the *mode*, different notes in the scale were used more than others. The note which was used most often in a particular mode was called the "dominant", and the note on which melodies ended was the "final". The dominant was often a perfect fifth above the final [Scholes 1970].

As triadic harmony developed during the Renaissance, the modal system was gradually supplanted by the diatonic system in which only two scales (major and minor) were theoretically recognized. The fifth relationship usually found between dominant and final (now called the *tonic*) became standard. The triad formed by adding a major or minor third between the tonic and dominant was called the *tonic triad.* The tonic note was regarded as the *root* of this triad.

The tonic triad acted as a kind of home base – a point of departure, and of return – in diatonic chord progressions. A common feature of chord progressions written from the seventeenth century on was the maintenance of a strong tonal centre by the use of chords harmonically related to the tonic, i.e. chords whose roots belong to the prevailing diatonic scale and are close to the tonic on the cycle of fifths. The resultant *functional harmony* [Riemann 1893] had a kind of stylistic unity and balance which had not been present in earlier musics.

The tonal framework set up by such a chord progression was called a *key.* Keys take the same name as their tonic triad (e.g. E. minor, G major). The key of a progression can change by a process called *modulation.* The chord or chord at the point of modulation (pivot chords) normally belong or are closely related to both preceding and following keys. The original key and tonic normally return at the end of a diatonic piece.

1.1.5 A Scientific Basis?

Until the start of this century, Western musicians generally believed that the diatonic system was in some way "natural". Helmholtz [1863] gave this belief a scientific basis through his investigations of the physics, physiology and psychophysics of tone perception, including the effect of the beating of near-coincident pure tone components on the dissonance of tone simultaneities, and the effect of frequencies in common on pitch relationships between sequential sounds. Helmholtz did not, however, believe that the diatonic system was an inevitable consequence of the psychophysics of tone perception. On the contrary, he emphasized the role of culture and aesthetics in the development of musical style.

Diatonic harmony dominated Western art music until the early twentieth century. Gradually it became clear that most of the possibilities of the system had been exhausted. Composers, seeing the need to experiment with new styles, began to regard physically based arguments in favour of the diatonic system as inadequate or spurious. Consonance and dissonance, as investigated by Helmholtz, appeared to be specific to Western music. Inflexible approaches to consonance and dissonance had inhibited the progress of music on previous occasions – for example, unprepared seventh chords were not accepted until Monteverdi because of theoretical considerations – and composers such as Schoenberg did not want such theoretical ideas to prevent progress into new harmonic styles [Griffiths 1978].

Nondiatonic styles met with mixed success. The fact that many pieces have now been accepted into the concert repertoire supports the idea that any new style can be appreciated given sufficient exposure to that style, and provides evidence against the theory that diatonicism is in some way more "natural" than other ways of organizing pitch in music. But most atonal (or twelve-tone) styles have still not been absorbed into the mainstream of Western music – perhaps because of the high rate of information processing required of the listener [Klingenberg 1974]. Meanwhile, "old-fashioned" tonal styles continue to flourish. If the success of innovations is gauged by their long-term popularity, then the greatest progress in tonal structure this century has occurred in the genres of impressionism and jazz.

Composers of twelve-tone music generally agree that most of the procedures they use to organize their music are imperceptible, even to trained listeners [Lerdahl and Jackendoff 1983, p. 299], and musicologists and psychologists agree that twelve-tone music has no sensory basis [Bruner 1984]. At the time of this study, however, there is still widespread disagreement concerning what, if any, sensory ("natural"?) basis exists for diatonic harmony. The rest of this chapter is concerned with attempts to find such a basis.

1.2 Physically Based Theories

1.2.1 Introduction

Perhaps the most obvious place to look for a scientific explanation of music theory is in the "real" worlds of physics and biology: the acoustics of sound transmission, the physiology of sound detection by the ear and the neurophysiology of sound "interpretation" in the brain. A physically oriented approach lacks many of the conceptual problems of psychological approaches. More importantly, physical measurements are made in well-defined ways, and tend to be easier to control and to reproduce than more subjective, psychological measurements.

The following aspects of the physics of sound (as explained more fully in [Wood 1961; Backus 1969; Benade 1976; Roederer 1979; Hall 1980; Moore 1982; Rossing 1982; Pierce 1983]) seem relevant for the theory of musical intervals and harmony. (i) The *waveform* of a sound (a finite-duration function of air pressure against time) can be analyzed into sinusoidal (pure tone) *components* by spectral analysis; (ii) pairs of pure tones in simple, whole-number *frequency ratios* can be combined (by addition of their waveforms) to form *periodic* waveforms; (iii) pairs of periodic waveforms combine to form amplitude-modulated waveforms (*beats*) if their frequencies are close, but not identical; and (iv) pairs of tone components in distorting electroacoustical systems produce *combination tones*. At first sight, the above phenomena appear to explain aspects of music perception such as the perception of (i) pitch, (ii) intervals, (iii) dissonance, and (iv) the roots of chords.

1.2.2 Frequency Ratios

It has been held since Pythagoras that certain musical intervals are simple or basic because they correspond to simple frequency ratios. The "harmony of the spheres" concept [Schneider 1960; Schavernoch 1981; Haase 1986], originally developed by Pythagoras and Plato [Lippman 1963a], persisted through the Middle Ages (e.g. Boethius, around 500 A.D.) and into the Renaissance (e.g. Zarlino, see [Wienpahl 1959]). A more "naturalistic" approach to frequency ratios was adopted by Rameau [1721] (Sect. 1.4.1). Modern music theorists such as Boomsliter and Creel [1961], Tanner [1981] and Barlow [1987] appealed to psychology and neuropsychology to explain the supposed sensitivity of the auditory system to frequency ratios.

The frequency ratio concept has some serious shortcomings. Not the least of these is that the very name "interval", as used by musicians, implies a kind of *distance*, not a ratio. Musicians assume that the same interval covers the same pitch distance, independent of transposition into different registers.

In the frequency ratio approach, most intervals may be assigned two different ratios, one of which contains only powers of two and three (the *Pythagorean* version) and one of which may also include powers of five (the *just* or *pure* version). For example, the interval of a major second (two semitones) may be assigned the frequency ratios 9/8 ($3^2/2^3$) and 10/9 ($2\times5/3^2$). If intervals corresponded to specific frequency ratios, one would expect intonation to vary randomly about one of these. In fact, intonation of intervals with different just and Pythagorean versions normally varies over a range which encompasses both ratios [Burns and Ward 1982], making it impossible to establish experimentally which of these is "the" frequency ratio of the interval.

If intervals were perceived as frequency ratios, then different *enharmonic spellings* of the same interval might be expected to correspond to different ratios. Consider, for example, the augmented fifth and minor sixth. These intervals are enharmonically equivalent: both span eight semitones in the chromatic scale. The minor sixth may be regarded as an octave (2:1) minus a major third (5:4), so its frequency ratio should be $(2/1)/(5/4) = 8/5 = 1.60$. An augmented fifth may be regarded as two superposed major thirds: $(5/4)\times(5/4) = 25/16 = 1.56$. In the frequency ratio approach, then, an augmented fifth is smaller than a minor sixth. Musical experience suggests exactly the opposite: the upper note of an augmented fifth interval normally resolves by rising a semitone, and tends to be played slightly sharp (if the instrument allows) to emphasize its "leading" nature, while the upper note of a minor sixth normally resolves by falling a semitone, and is therefore more likely to be played flat than sharp. Similarly, augmented fourths are consistently played "sharper" than diminished fifths [Ward 1970], even though the simplest (just) frequency ratio of an augmented fourth is smaller than that of diminished fifth.

The frequency ratio theory leads to the idea that harmonic language could be expanded by introducing frequency ratios including factors not only of 2, 3 and 5, but also of 7, 11, 13, etc. [e.g. Fokker 1966]. But in spite of decades of experimentation, Western harmony shows no inclination to move in this direction (Sect. 3.3.2).

1.2.3 Harmonic Series

Music theory texts generally make some reference to the harmonic series: the series of frequencies of harmonics of a single musical note. The first sixteen harmonics of the note C_2 are notated musically in Fig. 1.1d. Of these, the seventh (B♭), eleventh (F♯) and thirteenth (A♭) are noticeably out of tune by comparison to equal temperament.

Since Rameau [1721], the fourth, fifth and sixth harmonics have been regarded by many as the origin of the major triad, expressed in frequency ratio form as 4:5:6. The harmonic series seems to explain why we call the triad C–E–G a "C chord" as opposed to an "E chord" or a "G chord": the root of the chord is octave equivalent with the fundamental of the harmonic series from which it is supposed to be derived.

The harmonic series idea begins to break down as soon as we try to "explain" the second most common chord in Western music, the minor triad. The simplest conceivable frequency ratio representation of the minor triad is 10:12:15 (6:7:9 is out of tune). To suggest that the minor triad is derived from the 10th, 12th and 15th harmonics of a complex tone is far-fetched, on three counts. (i) Chords such as the diminished triad (5:6:7?) should, by this argument, be more consonant than the minor triad (10:12:15). (ii) The tenth harmonic is not octave-equivalent with the first, so the harmonic series model does not predict the root of the minor triad in the direct way that it predicts the root of the major triad. (iii) Harmonics above the tenth are generally inaudible in typical complex tones [Plomp 1964; Terhardt 1979a], so it is hard to see how they could play any direct role in music.

The present study describes a more sophisticated version of the harmonic series model of harmony, based not on the "naturalness" of the harmonic series as a physical entity (the frequency relationship between harmonics of complex tones) but on the *familiarity* of the auditory system with the harmonic series, via the pitch pattern of the audible harmonics of complex tones. In later sections, it will be shown how this approach solves problems which have traditionally plagued the harmonic series approach, such as the problem of tuning and frequency ratios (Sect. 3.3.3) and the problem of "nonharmonic" notes in musical chords, such as the third of the minor triad (Sect. 3.4.2).

1.2.4 Beats

Helmholtz [1863] was aware of the arbitrary nature of explanations of musical intervals and chords based on frequency ratios and the harmonic series. He developed an alternative scientific explanation of musical intervals in terms of *beats*: simultaneous pairs of periodic complex tones in simple frequency ratios sound consonant because some pairs of harmonics coincide and combine. In the case of the perfect fifth, for example, the third harmonic of the lower tone coincides with the second harmonic of the higher tone. If a musically important interval like a fifth is mistuned by a small amount, then pairs of near-coincident harmonics produce dissonant beats.

The Helmholtz theory of beats only directly applies to *simultaneous* tones. According to the theory, the musical meaning of intervals between *sequential* tones is somehow learned by exposure to simultaneous tones. History suggests, however, that musical intervals (octaves, fifths) between sequential tones existed long before people started singing or playing tones simultaneously in music [Grout 1960]. Further, melodies in which the most important notes are separated by octave, fifth and fourth intervals have evolved independently in different cultures [Burns and Ward 1982]. Wood [1961] points out that many melodies were originally sung in caves, where reverberation may have caused beats to influence interval sizes. However, an echo generally has much less physical amplitude than the original signal, so the amplitude modulation (beating) of near-coincident harmonics in the original and the echo would normally be imperceptible [Terhardt 1974b]. Thus, the theory of beats does not seem to be able to explain the melodic origins of musical intervals. (A more satisfactory explanation involves the *pitch commonality* of sequential tones, see Sect. 3.2.3.)

Beat frequencies exceeding 20 Hz evoke a sensation known as *roughness*, which is normally perceived as dissonant. The roughness of a chord does not indicate what musical intervals lie between the notes of the chord; it depends as much on the spectra and registers of the notes as it depends on the intervals between them. So it is questionable whether beats and roughness have anything to do with the "musical meaning" of simultaneous intervals and chords [Pierce 1966; Terhardt 1974a].

1.2.5 Combination Tones

Flute and recorder players are sometimes disturbed by low, buzzing sounds which seem to accompany their duet music. These "combination tones" result from non-linearities in the ear's frequency-encoding system (the cochlea and basilar membrane). The frequencies of combination tones depend on those of the actual pure tone components which produce them. One kind of combination tone is the *simple difference tone.* Its frequency is simply the difference between the frequencies of the two other tones. If these two tones are adjacent harmonics of a fundamental frequency, then the frequency of the simple difference tone is the same as that of the fundamental. A more commonly audible kind of combination tone is the *cubic difference tone*, equal to twice the lower frequency minus the higher frequency [Moore 1982]. For example, pure tone components at frequencies of 200 and 300 Hz produce a cubic difference tone at $2 \times 200 - 300 = 100$ Hz.

The pitch of a *residue tone* (a complex tone with no fundamental) normally corresponds to that of its "missing fundamental". Examples of residue tones in the modern environment are speech vowels heard over the telephone and musical tones produced by a pocket radio. Helmholtz [1863], and many of his successors up to and including Fletcher [1924], thought that the pitch of residue tones was due to combination tones, i.e. non-linear distortions in the cochlea. However, Schouten [1938] showed experimentally that the pitch at the "missing fundamental" is perceived even when no such distortion products are present.

Music theorists such as Tartini [1754], Krueger [1910] and Hindemith [1940] argued that the root of a chord (Sect. 1.1.2) is a combination tone. This theory was discredited by Stumpf [1909] and Cazden [1954], who pointed out that combination tones in music are generally inaudible (due to masking, see [Plomp 1965; Smoorenburg 1972], and, when they *are* audible, they sound dissonant (whereas chords with clear roots are consonant).

In summary, theories of musical intervals based directly on frequency ratios, the harmonic series and combination tones are implausible and unscientific. In the words of Cazden [1954, p. 289], "These doctrines represent in reality a species of mysticism dealing with magic numbers."

1.2.6 Periodicity

The above theories all take as their starting point the frequency spectrum of a sound, derived from its waveform by spectral analysis. Could an approach based directly on the waveform of a musical sound yield more feasible explanations of musical intervals?

The frequency spectrum of an exactly periodic waveform is exactly harmonic, i.e. its component frequencies are exact whole multiples of the fundamental (or waveform) frequency. So detection of harmonicity in the frequency domain (i.e. detection of a harmonic series of frequencies) is essentially equivalent to detection of periodicity in the time domain. Perhaps, then, the pitch of residue tones and the "meaning" of musical intervals can be explained in terms of the periodicity of their pressure waveforms, instead of the harmonicity of their frequency spectra. (For explanatory notes and illustrations, see the references cited in Sect. 1.2.1.)

The ear itself is not physiologically suited to periodicity detection. Theories of periodicity detection in hearing normally refer instead to the physiology of the brain. A starting point for such theories was the finding that neuron firing in the auditory nerve can, under certain conditions, by synchronized with the periodicity of the waveform of a tone [Wever and Bray 1937; Rose et al. 1967].

A possible explanation for the pitch of a residue tone based on periodicity is that neural signals from different parts of the basilar membrane, corresponding to different harmonics of the tone, are synchronized such that the combined signal (in the auditory nerve) has a periodicity corresponding to the frequency of the missing fundamental. Explanations of pitch based on periodicity of nerve impulse patterns date back to the *residue* theory of Schouten [1940]. The "musical meaning" of intervals corresponding to exact frequency ratios may be explained in a similar fashion [Boomsliter and Creel 1961]. The theory may be further refined by allowing for limited departures from periodicity to account for pitch shifts, octave stretch, and the pitch of slightly non-harmonic complex tones [Ohgushi 1983].

A theory of musical intervals based on periodicity of neuron firing is fraught with technical and philosophical problems. For the moment at least, it is impossible to establish a definite link between the sensation of pitch produced by a residue tone and the synchrony of the signals in the auditory nerve corresponding to its harmonics: any given sensation (or, for that matter, any experience) could be linked to any of countless simultaneous neural processes (Sect. 2.1.2). Until we understand some basic things about brain function, such as exactly how large ensembles of neurons behave under certain conditions and why, musically useful neural explanations of music perception will remain unattainable.

1.3 Psychologically Based Theories

1.3.1 Introduction

Purely physically based theories of harmony would have us believe that music is a physical phenomenon, heard by a frequency-analyzing ear connected to a neural network which is sensitive to periodicities of neural pulses. In a psychological approach, by contrast, music is inherently social; it is heard and appreciated by people, not organisms.

In music theory and analysis, there is little coordination between the activity of analyzing the score and that of analyzing the *experience* of the music [Clifton 1983]. Modern psychological models attack this problem by concentrating on *perceptual* (rather than physical) interactions between tones and groups of tones (chords, melodies, progressions, key areas, etc.). The cognitive approach to musical pitch perception, for example, looks for "structural relations within a set of perceived pitches independently of the correspondence that these structural relations bear to physical variables" [Shepard 1982, p. 306]. The same may be said for the linguistically based approach of generative grammars [Lerdahl and Jackendoff 1983] and the mathematically based approach of group theory [Balzano 1980]. These three modern, psychologically based approaches to harmony theory are briefly introduced in the following sections.

1.3.2 Cognitive Structures

In a cognitive approach, structural relationships between musical elements (tones, chords, keys ...) may be represented by points in a multidimensional space, by means of the mathematical technique of *multidimensional scaling* [Kruskal 1964]. The input data to the multidimensional scaling algorithm consists of a matrix of experimental results, in which each element in a set is compared with each other element (e.g. for apparent similarity). The output of the algorithm is a multidimensional map of the elements in the set, in which the distances between elements are approximately proportional to their apparent "differentnesses".

Shepard [1964] presented listeners with sequential pairs of octave-spaced tones: complex tones whose pure tone components are spaced at octave intervals over most of the audible spectrum. The listeners' task was to decide which of the two complex tones was higher in pitch. A multidimensional scaling solution of the results [Shepard 1978] yielded an almost perfectly circular solution, corresponding to the *chroma cycle* in music theory (a construct where successive pitch classes C, C#, D, etc. are substituted for numbers on a clock face). This result inspired a revival of the helical model of pitch, proposed as early as 1846 by Drobisch [Ruckmick 1929], in which tones an octave apart are represented by corresponding points on successive curves of a helix. Recently, more complex structures have been developed to account for musical pitch relationships in greater detail [Shepard 1982].

Krumhansl and Shepard [1979] asked listeners to rate how well a major scale was completed by an octave-spaced probe tone, presenting probe tones at all 12 different chroma during the experiment. The result was a *tone profile* reflecting the tonal structure of a major key. Tone profiles may be used to calculate the strength of the relationship between musical keys. For example, the profile for C major shows high positive correlations with profiles for closely related keys such as G major and A minor, and high negative correlations with distant keys such as F# major and G# minor. Multidimensional scaling solutions of the 24 by 24 matrix of correlation coefficients between musical keys yield maps in which closely related keys in music theory (dominant, subdominant, relative minor, parallel minor, etc.) are close to each other [Krumhansl and Kessler 1982].

Multidimensional scaling solutions of experimental data primarily *summarize* that data in a convenient and attractive form. They do not necessarily *explain* it. For example, the multidimensional scaling approach to the analysis of pitch relationships in music does not distinguish between universal (or sensory) and cultural effects. So it is not clear from multidimensional scaling solutions of similarity ratings of musical chords [Bharucha and Krumhansl 1983] that the data are largely sensory in origin (Sect. 5.7.3). The same applies for tone profiles of musical keys [Krumhansl and Kessler 1982] (Sect. 6.2.5).

Krumhansl and Shepard regard psychoacoustical data as essentially irrelevant to musical relationships. For example, Shepard [1982, p. 307] states that "because of the unidimensionality of scales of pitch such as the mel scale, perceived similarity must decrease monotonically with increasing separation between tones on the scale. There is, therefore, no provision for the possibility than tones separated by a particularly significant interval, such as the octave, may be perceived as having more in common that tones separated by a somewhat smaller but musically less significant interval, such as the major seventh." These criticisms of the psychoacoustical approach are invalid on two counts. First, the mel scale applies only to pure tones, not to musical (complex) tones (Sect. 2.5.2). Second, in a psychoacoustical approach, musical (complex) tones an octave apart have pitches in common: a musical tone has several possible pitches, separated by musical intervals such as octaves and fifths [Terhardt 1972, 1974a]. The latter explains why the similarity of sequential complex tones does not decrease monotonically with increasing pitch separation, but has local maxima at musically important intervals (Sects. 5.4–5.6).

The sensory approach of Terhardt and the cognitive approach of Krumhansl and Shepard are not mutually exclusive. They may be synthesized into a single method, by applying correlation and multidimensional scaling techniques to tone profiles calculated by psychoacoustical modelling (Sect. 6.2.5). In this way, aspects of harmony theory may be explained purely psychoacoustically, without the need to postulate the existence of abstract internal representations of musical structure.

1.3.3 Generative Grammars

Perceived music may be regarded as a *hierarchical structure of perceptual elements* [Deutsch and Feroe 1981; Krumhansl and Castellano 1983; Kessler et al. 1984]. Elements at one level of the hierarchy which are perceived to be related or similar are grouped to form elements at a higher level. The functions of chords in a musical key may be assigned to a hierarchy in which the tonic triad is at the top, diatonic chords are near the top, and chromatic chords are further down. The rhythmic functions of different metrical positions (bars, downbeats, off-beats, etc.) are similarly hierarchical. Schenker's [1935] concepts of *foreground*, *middleground* and *background* are examples of hierarchical levels applied to the analysis of entire musical pieces.

In their approach to perceptual hierarchies in music, Lerdahl and Jackendoff [1977, 1983] interpreted the conventions of music theory as a special case of the conventions of linguistic grammar. Linguistic conventions are intuitively understood by all speakers of a given language; musical conventions, by musically trained or experienced members of the corresponding musical culture. The theory of Lerdahl and Jackendoff provides a psycholinguistic basis for the communication of implicit or abstract ideas by music, while at the same time recognizing the pitfalls of making too close an analogy between music and language [cf. Feld 1974].

In general, a musical passage has a number of different possible hierarchical descriptions, corresponding to different analytical interpretations. These are illustrated by Lerdahl and Jackendoff by means of tree structures. Rules of "well-formedness" are used to decide which structural groupings are possible, and preferences for particular structural groupings over others are accounted for by means of "preference rules". Recent experimental work [Deliege 1987] has confirmed the validity of the grouping preference rules for musically trained listeners, and has shown that nonmusicians exhibit similar grouping behaviour.

Problems associated with the generative grammars approach to harmony include the failure to distinguish between compositional and perceptual structures, and the failure to account for the possibility of network structures and transient levels [Narmour 1983]. Furthermore, the theory does not explain *why* "... a triad in root position is more stable than its inversions; ... the relative stability of two chords can be determined by the relative closeness to the local tonic of their roots on the circle of fifths; conjunct linear connections are more stable than disjunct ones; ... and so forth" [Lerdahl and Jackendoff 1977, pp. 130–131]. In their book [1983], Lerdahl and Jackendoff discuss the problems associated with explanations of intervals, chords and scales based on the harmonic series, but provide no concrete alternative other than an appeal to cognition: "Tonality in music provides evidence for a cognitive organization with a logic all its own", and "musical idioms will tend to develop along lines that enable listeners to make use of their abilities to organize musical signals" (p. 293). This may be true, but it does not satisfactorily explain fundamental aspects of Western music such as the importance of the major and minor triads.

The methods of Lerdahl and Jackendoff are useful and practical teaching tools. The structure of tonal music becomes clearer to the student when it is expressed in terms of a strict formalism. Their work could be appropriately expanded by incorporating explanations of musical elements and preference rules based on Terhardt's [1974a, 1976] concept of pitch, consonance and harmony in music. The result would be a more comprehensive and satisfying theory of harmony than has previously been possible.

1.3.4 Mathematical Groups

Balzano [1980] contributed to the debate on the origins of the diatonic scales and triadically based harmony by an analysis of the mathematical properties of the set of twelve pitch classes in the chromatic scale. He treated this set as a cyclic group of order 12, i.e. as a group of 12 elements in which each element has an upper and a lower neighbour. This approach grew out of the application of set theory to the theory of atonal music [Forte 1964, 1973; Beach 1979]. Its most fundamental assumptions are that musical intervals are perceived categorically [Burns and Ward 1978] and that notes an octave apart are harmonically equivalent (Sect. 3.3.1).

Balzano generated the cyclic group of order 12 in various ways. The simplest method involves adding semitone intervals one at a time until all members of the group have been generated. The result is the chroma cycle $\{0, 1, 2, 3, \ldots, 11\}$, or "semitone group". Another method involves adding perfect fourth or fifth intervals, producing the cycle of fifths $\{0, 7, 2, 9, \ldots, 5\}$. The standard heptatonic scale consists of neighbouring elements in this set: 5, 0, 7, 2, 9, 4 and 11. No other intervals beside 1 (or 11) semitones and 7 (or 5) semitones can alone generate the entire cyclic group of order 12, a fact which Balzano interpreted as indicative of the importance of the semitone (or minor second) in melodic relationships and of the fourth and fifth in harmonic relationships.

A further method of generating the cyclic group of order 12 proposed by Balzano involved a two-dimensional generator. This idea was not entirely new. Schoenberg [1954] developed a two-dimensional map of musical keys, extending vertically by perfect fifths and horizontally by relative and parallel major/minor relationships, in which closely related keys were always close together. Longuet-Higgins [1979] developed a two-dimensional map of pitch classes, extending vertically by major thirds and horizontally by perfect fifths. Balzano's method was to express each musical interval as as a combination of major and minor thirds: in the terminology of set theory, he generated the cyclic group of order 12 by multiplying together cyclic groups of order 3 (there are 3 major thirds in an octave) and of order 4 (there are 4 minor thirds in an octave). The resultant two-dimensional space was similar to that of Longuet-Higgins in that the notes of common (major and minor) triads and of diatonic (major and minor) scales were next to each other, and tonic triads were centrally located relative to the notes of the diatonic scale.

In a later article, Balzano [1982] used the set-theoretical approach of Forte [1964] to demonstrate the uniqueness of the standard heptatonic scale. First, he defined a *coherent* scale to be one in which interval size increases monotonically with the number of scale steps, regardless of where one begins in the chroma cycle. Of the 66 different possible heptatonic scales (7-note subsets of the chromatic scale, not counting transpositions), only the standard heptatonic scale is coherent.

Balzano's theory has the following drawbacks. First, it ignores the sensory basis for the musical importance of the perfect fifth and fourth intervals

[Terhardt 1974a, 1976]. In order to "derive" the importance of those intervals, Balzano presupposes the existence of the chromatic scale; but this scale first emerged in Western music long after fifths and fourths came into widespread use (Sect. 3.3.2). Second, it fails to give a detailed account of the *process* by which the described group-theoretical properties may have influenced the historical development of music. Consequently, the *extent* of this influence is also unclear. Finally, the group-theoretical principles which Balzano invokes to demonstrate the usefulness of the 12-note scale also suggest divisions of the octave into 20, 30 or 42 equal parts [Balzano 1980]; but such scales have failed to generate viable new harmonic systems.

1.4 Towards a Psychophysical Theory

1.4.1 From Rameau to Terhardt

In his *Traité de l'harmonie* [1721], the composer and theorist Jean-Philippe Rameau adopted a physical entity, the *corps sonore*, as the basis for his theory of harmony [Christensen 1987]. By this he meant any vibrating system (string, pipe, vocal chords, etc.) whose sound spectrum comprises whole-number multiples of a fundamental frequency. Rameau hypothesized a link between the fundamental frequency of the *corps sonore* and the *basse fondamentale*, or root, of a chord. This led to a distinction between chords whose bass note is octave-equivalent to the root (root position chords such as G–B–D) and chords whose bass note and root are different (inversions such as B–D–G, D–G–B).

Rameau's seminal work on harmony inspired generations of music theorists, especially during the music-theoretical boom of nineteenth-century Europe [Watson 1982]. To this day, harmony textbooks begin with a discussion of the musical intervals of the harmonic series. But, as we have seen (Sect. 1.2.3), the harmonic series approach fails to explain satisfactorily such elementary structures as the minor triad and the diatonic and chromatic scales, in spite of valiant efforts to force these structures into the harmonic series mould by music theorists.

The trouble with Rameau's approach seems to be that the *corps sonore* is an objective, *physical* entity, but the musical harmony which it is supposed to explain is a subjective, psychological *experience*. The seeds for a resolution of this problem were sown early this century with the emergence of the scientific discipline of *psychophysics*, or – more specifically – *psychoacoustics*. Psychoacoustics is concerned with the relationship between physical and perceptual descriptions of sound, e.g. between frequency, intensity and spectral composition on the one hand, and pitch, loudness and timbre on the other [Fletcher 1934].

A turning point in the theory of harmony was Terhardt's [1974a, 1976] psychoacoustical reinterpretation of Rameau's theory. Instead of adopting a physical entity (the *corps sonore*, or, in modern parlance, the *harmonic complex tone*) as the ultimate basis of harmony, he adopted the principle of *familiarity of the auditory system with harmonic complex tones.* He hypothesized that our sensitivity to musical harmony and consonance is ultimately based on our sensitivity to the specific pattern of pitch which characterizes harmonic complex tones, a sensitivity which is fostered primarily by repeated exposure to this specific pattern in speech vowels, and which presumably influenced Western harmony at all stages of its development. Terhardt thus shifted the emphasis in Rameau's theory from the world of physics and "Nature" to the world of sensation and perception. In this way he was able to solve the main conceptual and (eventually) the main practical problems of Rameau's theory.

Rameau was not entirely oblivious of the role of perception in his theory, nor of the importance of speech vowels. He began his *Nouveau système de musique théorique* [1726] as follows:

> There is *actually in us* a germ of harmony which apparently has not been noticed until now. It is nonetheless easily perceived in a string or a pipe, etc. whose resonance produces three different sounds at once. Supposing this same effect in all sonorous bodies, one ought logically to suppose it in the *sound of our voice*, even if it is not evident. ([Christensen 1987, p. 26] my italics; the "three different sounds" to which Rameau refers are those of the octave, twelfth and seventeenth, i.e. of harmonics 2, 3 and 5.)

However, Rameau was working long before the major developments in the psychoacoustics of pitch perception of the nineteenth and twentieth centuries [Ohm 1843; Helmholtz 1863; Fletcher 1934; Schouten 1940; Plomp 1967; Terhardt 1972] and so was in no position to realize the implications of a perceptual approach to the relationship between harmony and the *corps sonore.* Nor was he in a position to identify the important role of the voice as the most important *corps sonore* underlying musical harmony (Sect. 3.1.2).

1.4.2 The Psychoacoustical Approach

The conventions of Western music theory are general statements about what "sounds right" in diatonic music. As we have seen, attempts to find a basis for music theory in *acoustics* run into difficulties because acoustics is based on physical measurements – what sound "really is", perhaps, rather than what it sounds like. Attempts to find a basis for music theory in *psychology* have so far run into difficulties because psychologists have tended to assume a simple one-to-one correspondence between the notes of a musical element (such as a chord or melody) and the tone sensations evoked by that element. This relationship is actually quite complex [Terhardt et al. 1982b].

Psychoacoustics is concerned with the relationship between the physical properties of stimuli and observers' descriptions of the sensations they evoke. On the basis of experimental results indicating the nature of this relationship in specific cases, general rules are developed by which sensory properties of sounds may be calculated from their acoustical properties. This general understanding allows the sensory properties of sounds, as distinct from their physical properties or musical notation, to be accurately described [Terhardt 1978].

Psychoacoustical models allow the sensory nature of sounds to be specified before higher psychological processes such as hierarchical grouping are considered. A complete analysis of music perception involves appropriate acoustical description of musical sounds, psychoacoustical theories and models by which acoustical descriptions of sounds may be converted into appropriate experiental descriptions, and appropriate psychological theories by which the organization of (and selective attention to) musical sounds may be understood.

Psychoacoustics is normally restricted to isolated sounds, i.e. sounds out of context. Extrapolation of results to sounds in a musical context can be problematic. In the present study, limited extrapolation of this kind is achieved by means of quantitative estimates of the "pitch commonality" and "pitch proximity" of sequential sounds (Sect. 4.6), bridging the gap between existing psychoacoustical and psychological approaches.

In psychoacoustics, experiential (or psychological) variables such as "loudness" and "roughness" are measured experimentally, usually by averaging the results of a number of different listeners. Such averaging is only meaningful when results vary relatively little from one person to the next. Care must be taken when transferring psychoacoustical results across linguistic or cultural boundaries, as translation and cultural conditioning may have a considerable effect on the measurement and scaling of psychoacoustical variables.

A disadvantage of the psychoacoustical approach is its tendency toward excessive complexity. For example, "appropriate evaluation and prediction of the various pitch percepts of complex tonal signals on the basis of the physical parameters is far from being either simple or trivial" [Terhardt et al. 1982b, p. 679]. The complexity of psychoacoustical modelling may be reduced by isolating effects which are relatively unimportant in a given application, and neglecting them (as in Chap. 4 of the present study).

1.4.3 Outline of the Book

This study aims to explain some of the conventions of Western music theory (as outlined earlier in this chapter), using psychoacoustical data, theories and techniques to explore their sensory basis. Chapters 2 and 3 survey relevant literature in psychoacoustics and psychomusicology, and introduce some new concepts and approaches to music theory. Chapter 4 describes a mathematical model by which sensory (or universal) aspects of the perception of chords and chord sequences may be simulated. The model is then tested and fine-tuned by comparing its calulations against the results of experiments (Chap. 5): deviations between calculations and experimental data point to specific effects of musical conditioning. Finally, the model is applied in the theory, analysis and composition of music (Chap. 6). Technical terms from music, psychology and physics are explained in an extensive glossary.

2. Psychoacoustics

Psychoacoustics investigates relationships between the physical properties of sounds (waveform, spectrum, level, frequency, ...) and the way sounds are experienced (loudness, pitch, timbre, salience). The first stage of auditory perception involves spectral analysis in the cochlea, with specific time and frequency characteristics. Thereafter, analytical information is extracted by categorical perception, and holistic information (which can ambiguous, depending on context) is extracted by pattern recognition. In a psychoacoustical approach, the perception of complex tones (and hence of ordinary environmental sound sources) involves the spontaneous recognition of harmonic patterns among the pitches of audible pure tone components. Consequently, the pitch of complex tones (and even of pure tones) can be ambiguous. Pitch may be measured and perceived on continuous scales (in psychoacoustics) and categorical scales (in music); the latter case includes the recognition of both intervals (relative pitch) and notes (perfect pitch) by musicians.

2.1 Philosophy of Perception

2.1.1 Hardware and Software

Within limits, it is useful to draw an analogy between the brain and the hardware of a computer. The way we perceive, by this analogy, is like a computer program – a software package for the brain [cf. Lilly 1974].

There is no sharp boundary between hardware and software in computing. A lot of what is called hardware is in some sense programmed to perform specific transformations on input signals. The same may be said for perception and behaviour.

The software of perception develops quite differently from contemporary computer software. It is acquired ("learned") as the organism actively explores and interacts with its environment. In this respect, the brain may be said to be *self-programming*. The program by which it programs itself is "innate" or "instinctive". The self-programming process involves interaction of the whole organism with its various environments; it begins before birth (Sect. 3.1.2), and continues throughout life.

Hardware and software can be remarkably independent of one another; the same computer can run completely different kinds of program (i.e. perform completely different algorithms), and the same program can be performed on completely different kinds of computer (e.g. serial versus parallel processors). Similarly, the nature of perception may be largely independent of the particular ways in which the human brain stores and processes information.

In particular, music perception does not necessarily depend on brain physiology; Roederer's suspicion [1987, p. 82] that "... 'universal' characteristics of music are ... the result of built-in physiological or neuropsychological functions of the auditory system" probably applies only to the physiology of the ear (e.g., its frequency analyzing property). Instead, the nature of music would appear to depend primarily on the way the auditory system interacts with sound, considered as a part of the interaction of the organism with its environment [Gibson 1979]. Most aspects of the perception of music may be satisfactorily explained in terms of familiarity with environmental and musical sounds (Sect. 3.1).

2.1.2 Matter, Experience and Information

A useful philosophical basis for the study of music perception is the *three worlds* concept of Karl Popper [Popper and Eccles 1977; Terhardt, personal communication]. World 1 is the world of *matter* (and energy): it comprises physical objects, states and processes, and includes musical instruments, tones, the ear and the brain. World 2 is the world of *experience,* or states of consciousness. It includes all aspects of musical experience – sensations of tone, harmony, rhythm, consonance and tonality, as well as the emotions evoked by a piece of music. The contents of world 3 may be variously described as symbols, descriptions, language, "objective knowledge", or simply *information.* World 3 includes thoughts and ideas, literature, computer programs, musical scores, and music theory.

The degree to which correspondences exist between the three worlds is limited; each world is, to some extent, autonomous. The limited correspondence between worlds 1 and 3 (matter and information) is reflected by Heisenberg's uncertainty principle in quantum mechanics – a special case of the general rule that you can't measure something without in some way changing what you are measuring. The limited correspondence between worlds 2 and 3 (experience and information) is reflected by the existence of "feelings which cannot be put into words". In the case of worlds 1 and 2 (matter and experience), brain states and associated experiences are measured and expressed in fundamentally different ways, involving physical measurements (expressed in physical units) on the one hand and observers' introspective reports (expressed in natural language) on the other.

There is no clear a priori justification for the belief that all aspects of experience may someday be predictable on the basis of physiological measurements, no matter how sophisticated such measurements might become in the future. In the words of Gibson [1979, p. 306],

"Perception cannot studied by the so-called psychophysical experiment if that refers to physical stimuli and corresponding mental sensations. The theory of psychophysical parallelism that assumes that the dimensions of consciousness are in correspondence with the dimensions of physics and that the equations of such correspondence can be established is an expression of Cartesian dualism. Perceivers are not aware of the dimensions of physics. They are aware of the dimensions of information in the flowing array of stimulation that are relevant to their lives."

Moore [1982] in his book aimed to specify the relationships between sounds and sensations "in terms of the underlying mechanisms", seeking to "understand how the auditory system works, at well as to look as what it does" (p.1). The phrase "underlying mechanism" betrays Moore's belief in concrete relationships between stimulus and sensation at the level of brain function. In the light of Gibson's comments (above), Moore may well be asking unanswerable questions.

It is widely believed that only the physical world *really* exists, and that physical states and processes underlie both experience and information. This raises some thorny questions. If experiences don't really exist, for example, what is the point of funding the arts? And if information does not really exist, exactly what *was* it that Mozart bequeathed to humanity? A contrasting (and equally valid) view is that experience is the foundation and final arbiter of knowledge [Clifton 1983]. According to this view, the existence of the physical world is just a hypothesis based on everyday experience of, and theories about, the environment. If this is the case, however, why is it that the physical world can be described and measured more precisely than the worlds of experience and information? In Popper's approach, philosophical problems such as these are avoided by regarding matter, experience and information as equally real.

Gödel's theory in mathematics may be interpreted to imply that no theory or philosophy can explain itself: all abstract systems incorporate inconsistencies [Hofstadter 1980]. Popper's three worlds concept is no exception. For example, a *thought* may be regarded as either a piece of information or an experience. On the other hand, all scientific research relies on some kind of paradigm [Kuhn 1962]. The three worlds concept is chosen as a paradigm on which to base a theory of music perception, not because it is perfect (it isn't), but because it clarifies the multidisciplinary mosaic of music perception research.

An example of the application of the concept in the case of music performance is Nakamura's [1987] study of the relationships between "the dynamics of a piece of music that professional performers intend to convey to listeners [world 3]; ... the intensity of tones produced by the performers [1]; and ... the listeners' perception of the dynamics of performances [2]" (abstract). Further examples are described below.

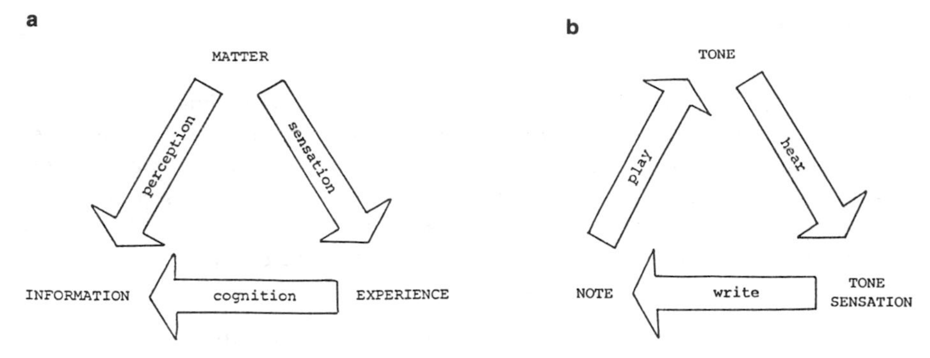

Fig. 2.1 a, b. Popper's three worlds and music perception. **a** Perception, sensation and cognition. **b** A cycle of musical creation

2.1.3 Perception, Sensation and Cognition

Perception is an active process by which organisms extract information from and interact with their environments [Gibson 1966]. Sensation, by contrast, is passive. It involves experiencing or being aware of sensory input, without necessarily focussing on environmental objects.

In traditional psychological paradigms, perception is regarded as a two-stage process involving the subprocesses sensation (as studied in psychophysics) and cognition (Fig. 2.1 a). In the first stage, physical stimuli are "converted" into sensations. In the second stage, hypotheses about the environment are made on the basis of available sensations; the results may be called *percepts* of environmental objects. In the traditional approach, then, sensation is regarded as an essential prerequisite for perception.

Gibson [1966] observed that most environmental interaction is almost entirely "automatic", occurring with little or no awareness of the analytic complexity of associated sensory patterns, cognitive processes and motor responses. This suggests that the perceptual extraction of information from the environment occurs much more directly than in the traditional two-stage model. Gibson consequently demoted sensations from their traditional status as prerequisites for perception to the more realistic status of mere *byproducts* of perception.

Perceptual theories may be divided into three kinds: those based on psychophysics (the interaction between worlds 1 and 2), cognition (2 and 3) and direct or ecological perception (1 and 3). In the present study, psychophysical and (to a lesser extent) direct or ecological explanations of music perception are generally preferred to cognitive ones. Psychophysical and direct perceptual explanations have the advantage that they involve the physical world directly, and the physical world is more experimentally measurable and precisely specifiable than the worlds of experience and knowledge. Because cognitive theories relate the "subjective" worlds of experience and knowledge to each other, they lack the stability of being anchored to "objective" physical measurements.

The music listener experiences both sensations and percepts. Musical percepts correspond to physically real objects like singers, instruments and loudspeakers. Musical sensations include tone sensations, and their pitch, loudness, and timbre. Rhythm, melody and harmony also fall under the heading of sensations, as they have no specific correlates in the everyday, non-musical environment.

2.1.4 Tone, Tone Sensation and Note

The word "tone" in this study refers to a *physical* entity: a periodic acoustical disturbance which can evoke pitch. A "note" is an *instruction* to play a tone. In addition, the term "tone sensation" is used to refer to the *experience* accompanying the perception of a tone.

In experimental acoustics, the basic measurement of a tone is its *pressure waveform*: a function of oscillatory pressure against time, recorded at some point in space by means of a microphone. The amplitude and phase spectra of a tone, obtained by Fourier (spectral) analysis of its waveform, may be used to recreate the original waveform by adding component waveforms.

Tone sensations (or "sensory tones" [Terhardt 1979b]) are defined to be experiences associated with the perception of tone sources, such as people speaking, musical instruments being played, and so on. Tone sensations have the attributes salience (perceptual importance), pitch, timbre, (apparent) onset time and (apparent) duration.

Notes belong to the world of information. The attributes of a note correspond not to the physical attributes of the tone to be played but to its *perceptual* attributes, expressed by means of labelled *categories* (C#, quarter note, etc.). Pitch categories in Western music (Sects. 2.5.3–2.5.5) are normally specified relative to each other and to tonally important pitch categories by means of interval categories expressed in semitones. Time categories, used to express both the onset time and duration of a note, are specified as rational numbers (fractions of whole numbers) relative to a prevailing pulse or metre. In performance, the actual sizes of notated pitch and time intervals depend on performer and context [Sundberg 1982; Gabrielsson et al. 1983; Sundberg et al. 1983]. Different performances of the same music notation nevertheless remain within certain limits (category boundaries) implied by the notation.

In Western music, loudness and timbre are notated separately from pitch, onset time and duration. Loudness is indicated categorically by *dynamics* corresponding to ordinary words such as "loud", "very soft", etc. Timbre is indicated categorically in music notation by *orchestration*: the names of musical instruments, and instructions for the use of mutes, special techniques, and effects such as pizzicato and flutter tonguing. The actual (experienced) loudness and timbre and corresponding physical characteristics of a tone played according to a particular musical score and on a particular instrument depend on pitch, context, player and so on.

The relationship between tones, tone sensations and notes may be regarded as a cycle of musical creation which links the performers, audiences and composers of Western music (Fig. 2.1 b). Composers write instructions to performers in the form of scores. In sight-reading, the performer sees a musical note, "thinks" it, and then plays it; the result is called a tone. The performance of musical tones is controlled by a kind of feedback mechanism by which the performer hears what has been played, checks whether its sensory attributes correspond to those required by the notation, the performer's concept of the music, and expectations of a real or imagined audience, and then makes appropriate adjustments to the performance.

In the case of improvisation (e.g. in the Baroque period, and in jazz), the word "write" in the figure may be replaced by "decide". Improvisers decide which notes to play on the basis of the kinds of sounds they have just created, and the direction they wish to take in the music. Similarly, the word "note" in the figure may be interpreted as a kind of self-instruction on the part of an improviser, referring (in a rather analytical way) to decisions made and executed during improvisation.

Decisions made during musical improvisation need not be conscious, and experienced note-readers are not necessarily conscious of the individual notes of a score as it is being sight-read. The idea of unconscious decisions, regulated partly by tone sensations experienced during a musical performance, may be used to explain the kind of feedback mechanism by which both music readers and improvisers control their performances. An account of psychological aspects of music reading and improvisation is given by Sloboda [1985].

From the physicist's point of view, tones are more important than the tone sensations which they evoke and the notes by which they are played. From the psychologist's point of view, tone sensations are basic, for without them we would never have developed the concepts of "tone" or "note". From the musician's point of view, notes are basic, because without notes no musical tones would be played and no musical tone sensations heard; even improvisers may be thought to imagine notes before playing them. These three views are internally coherent, but nonetheless limiting. It is preferable to assign tones, tone sensations and notes equal importance in an objective analysis of music perception.

2.2 Auditory Sensation

2.2.1 Loudness and Timbre

The sensory attributes of tonal sounds (i.e. simultaneities) most commonly investigated in psychoacoustics are loudness, pitch, and timbre. In the case of individual tones perceived within tonal simultaneities, one speaks in psychoacoustics of *salience* (sensory importance) rather than loudness. Like all psychoacoustical parameters, the sensory attributes of tonal sounds depend on listener and context.

The loudness, pitch and timbre of an isolated tone *all* depend on *all* the corresponding physical parameters: intensity, frequency and spectral composition (as a function of time) of the tone [Fletcher 1934]. So any physical change in a sound is likely to produce a change in all its sensory attributes. For example, changing frequencies of the pure tone components of a sound changes its loudness and its timbre.

Of all the sensory attributes of tonal sounds, pitch is the most important for harmony, and is dealt with in detail in later sections. Loudness and timbre are not so important for harmony, and are discussed briefly below.

The (subjective) loudness of a pure tone depends on its frequency as well as its sound pressure level [Fletcher and Munson 1933]. The loudness of a complex tone or sound also depends on its spectral distribution [Zwicker 1960] (Sect. 4.3.1). Loudness is measured in psychoacoustics by comparing the loudness of a test sound with that of a standard reference tone [American Standards Association 1960; Zwicker 1982]. The *loudness level* in *phon* of a sound is defined as the sound pressure level (SPL) in decibels (dB) of a (standard) 1 kHz pure tone when the sound and the standard tone are judged to be equally loud. For example, a sound which is just as loud as a pure tone of frequency 1 kHz and SPL 60 dB has a loudness level of 60 phon. Loudness level is an accurate, but not a proportional, measure of loudness. Doubling the (apparent) loudness of a sound doesn't double its loudness level, but increases it by about 10 phon. The corresponding proportional scale is called simply *loudness*, and is measured in *sone*, such that a (test) sound of loudness n sone is judged to be n times louder than a pure tone of frequency 1 kHz and SPL 40 dB (the standard). A loudness of 1 sone corresponds to a loudness level of 40 phon, 2 sone corresponds (approximately) to 50 phon, 4 sone to 60 phon, 8 sone to 70 phon, and so on.

Timbre (tone quality) is associated with the identification of environmental sound sources [Bregman and Pinker 1978], including musical instruments [Saldanha and Corso 1964]. Like vowel quality, timbre depends on the absolute frequencies and amplitudes of pure tone components. In addition, the physical characteristics of the onset of a musical tone are crucial for timbre and instrument identification [Berger 1964]. A powerful technique for the understanding of timbre is analysis-by-synthesis [Risset 1978]. Like pitch (tonality) and loudness (dynamics), timbre can be used to delineate musical forms in contemporary styles [McAdams and Saariaho 1985].

Timbre is multidimensional [Wedin and Goude 1972]. It may be quantified on various sensory scales such as "brightness" and "richness", and studied by multidimensional scaling of similarity ratings [Grey 1977]. Sensory dimensions of timbre which are important for the theory of harmony are *roughness*, associated with beating between pure tone components, and *tonalness*, the degree to which a sound has the sensory properties of a single complex tone such as a speech vowel [Terhardt 1983] (Sect. 3.2.2). Bad musical intonation (tuning) causes roughness to increase and tonalness to decrease; this explains the finding of Madsen and Geringer [1981] that deliberate mistuning in flute/oboe duets is often misinterpreted by listeners as bad tone production on the part of the performers.

2.2.2 Spectral Analysis

According to Fourier's theorem in mathematics, any waveform of finite duration (not necessarily periodic) may be expressed as a sum of component waveforms which are sinusoidal over the same duration. In acoustical terms, this means that any sound may be expressed as a sum of pure tone components. Note that these components are not directly measurable, so – strictly – they do not exist as physical entities. Instead, they are found by subjecting the waveform of a sound to a *mathematical* procedure: spectral analysis.

The relationship between sound input to the ear and the information conveyed to the brain is essentially the same as the relationship between a sound and its pure tone components. In this sense, the ear subjects incoming sounds to spectral analysis [Ohm 1843; Helmholtz 1863; Terhardt 1972, 1985]. This may be regarded as an early stage in the extraction of information from sound, in order to enable and facilitate interaction with the environment.

The cochlea is a bony, snail-like hollow in the petrous bone. The basilar membrane, which it houses, may be regarded as the receptor surface of the peripheral auditory nervous system. The basilar membrane is tapered: broad at one end and narrow at the other. When a pure tone is detected, waves travel along the membrane, reaching maximum amplitude at a point depending on the frequency of the tone [Bekesy 1947]. This spectral information is maintained in the peripheral nervous system [Evans 1975].

The importance of place on the basilar membrane in determining the pitch of pure tone sensations is supported by work with the partially deaf. Damage to part of the basilar membrane can cause deafness in a corresponding frequency range [Crowe et al. 1934], and electrodes implanted at different places in the auditory nerve of a deaf person produce tone sensations of different pitch [Simmons et al. 1965]. However, some experimental pitch data cannot be accounted for by place alone: it appears that both place information (e.g. which parts of the basilar membrane experience maximum displacement) and temporal information (e.g. the rate at which a particular part of the membrane oscillates) contribute to the pitch of pure tone components [Moore, 1982]. For example, below about 50 Hz, the position of maximum amplitude is independent of frequency [Bekesy 1947]. In this region, the pitch of a pure tone may depend on the rate of neuron firing in the auditory nerve [cf. Wever and Bray 1937]. In any case, Ohm's acoustical law, as described above, holds regardless of how the complex motion of the basilar membrane is translated into the pitch of pure tone components.

Like any spectral analysis system, the ear has limited frequency resolution. Simultaneous pure tone components must differ in frequency by a certain minimum amount before they can be resolved (or discriminated). Such minimum frequency differences are determined by the effective time constants (i.e. effective durations of the analysis interval) of the ear, which vary as a function of frequency [Terhardt 1985]. Simultaneous pure tones must be at least 1.0 – 1.5 semitones apart (considerably more than this at low frequencies) to be resolved, i.e. to produce distinct tone sensations [Plomp 1964; Terhardt 1968a].

The ear is not a perfect spectrum analyzer. Under certain spectral conditions, single pure tones of high sound level can produce harmonic distortion [Egan and Klumpp 1951], and simultaneous pure tones can produce combination tones (Sect. 1.2.5).

The output of the ear's spectral analysis is influenced mainly by that part of the incoming waveform immediately preceding the time of observation; earlier and earlier parts of the waveform influence perception less and less. An appropriate mathematical procedure for modelling this kind of spectral analysis is the *Fourier-t-transform* (or FTT), in which sound waveforms are multiplied by an exponential decay function, and spectral analysis is subsequently performed on a window extending from negative infinity to the present. In psychoacoustical applications, the variable amplitude and frequency dependencies of the FTT may be adjusted to fit those of the auditory system [Terhardt 1985]. When this is done, only audible pure tone components are output by the procedure, i.e. masking is automatically accounted for.

The *masked threshold* (or audiogram) of a pure tone is a graph of the sound pressure level (SPL) of a second, simultaneous, barely audible pure tone, as a function of its frequency [Wegel and Lane 1924]. It is roughly triangular in shape, peaking at the frequency and amplitude of the first tone: the closer the second tone lies to the first in frequency, the more it is masked, so the higher its SPL needs to be before it can be heard. For pure tones above about 500 Hz (C_5), the gradient of the lower-frequency side of the masked threshold is constant at roughly 9 dB per semitone.

As a rule, a change can be heard in a sound if part of its masked threshold undergoes a vertical shift of 0.5 – 1.0 dB [Riesz 1928; Zwicker 1970]. This implies that a change can be heard in a pure tone if it is shifted in frequency by 0.06 – 0.12 semitone. Difference thresholds of frequency as low as 0.02 semitone for the best listeners under ideal conditions [König 1957; Fastl and Hesse 1984] may be due to the added role of temporal information in pitch perception. Alternatively, they may be explicable in terms of musical experience: small pitch changes are more important in music than small loudness changes, and discrimination improves with practice [E. J. Gibson 1953].

2.2.3 Sensory Memory

Sensory memory is spontaneous memory, i.e. memory in the absence of attention, noticing, categorization, abstraction, semantic processing, etc. In a sense, this is not memory at all – it is a kind of spontaneous decay characteristic of the sensory system for which "memory" is the conventional psychological metaphor. To measure the duration of sensory memory it is necessary to ensure that a stimulus remains unnoticed for a specified time after its real-time occurrence.

The duration of visual sensory memory is about 0.1 – 0.2 s [Averbach and Coriell 1961]. Decay times in this range are also characteristic of forward masking effects (masking between sequential sounds) in psychoacoustics [Moore 1982; Zwicker 1982]. Auditory sensory memory, otherwise known as echoic memory [Neisser 1967] or precategorical acoustic storage [Crowder and Morton 1969] lasts much longer than both visual sensory memory and acoustical masking effects. Eriksen and Johnson [1964] estimated its duration at 10 s. Later researchers reported lower values such as 5 s [Glucksberg and Cowen 1970] and 2 s [Crowder 1970], suggesting that Eriksen and Johnson's experiment was influenced by ordinary, non-sensory memory.

Sensory memory linkage may be regarded as an essential prerequisite for the spontaneous perception of pitch relationships between sequential sounds (pitch commonality and proximity, Sect. 3.2.3). This is no problem in music, as the chords that make up chord progressions are normally much less than 2 s apart. In experiments to investigate pitch relationships (Chap. 5), the pairs of sounds presented in each trial followed each other at time intervals much shorter than 2 s. On the other hand, the time intervals between different trials in the experiments generally exceeded 2 s, so that sensory interference between trials was unlikely to affect results.

The duration of auditory memory increases considerably if sounds are noticed as they occur in real time. Sensory material persists longer in memory the more it is "processed through semantic levels" [Craik and Lockhart 1972], i.e. the higher it is abstracted in a perceptual hierarchy. Wickelgren [1969] found evidence of both sensory or "short-term" and categorical or "intermediate-term" memory for pitch, with durations of about 2 and 20 s respectively.

Memory for a particular sound is disrupted by intervening sounds [Wickelgren 1966; Massaro 1970; Deutsch 1972a; Dewar et al. 1977; Olsen and Hanson 1977]. Duration of memory for tones in an unfamiliar musical context tends to fall as the apparent rate of sensory information in that context increases. These effects are neglected in the present study, which is mainly concerned with sensory auditory memory in the absence of interference.

2.3 Extraction of Information

2.3.1 Noticing and Salience

To notice something is to become aware or conscious of it. This often involves assigning a verbal label to it. There is a large grey area between "noticed" and "unnoticed", in which objects and stimuli influence experience and environmental interaction, but are not necessarily assigned verbal labels.

In this study, the salience of an environmental object or stimulus is defined quantitatively as the probability that it will be noticed. In other words, the salience of the corresponding percept or sensation is its probability of occurring. If a sensation or percept already exists, then its salience may be regarded as a measure of its apparent importance or strength. For example, a chord may evoke several tone sensations, but some may sound more important than others.

The pure tone components of a complex tone are seldom directly noticed, yet each contributes to the perception of the tone as a whole (Sect. 2.4.3). The degree to which each contributes depends on its salience (Sect. 4.4.2). Similarly, the degree to which (unnoticed) tone components contribute to the strength of sequential pitch relationships depends on their salience (Sects. 4.6.1, 4.6.2).

Relatively salient tone sensations in a musical chord normally correspond to actual tones, and are recognized as such by musicians. Tone sensations with low salience do not normally correspond to actual tones, but to implied or harmonically related pitches such as the root of a chord in inversion (Sect. 6.1.5).

2.3.2 Categorical Perception

Categorical perception refers to the division of a perceptual continuum into labelled categories, specified by their centres and widths, or by the positions of their boundaries. Categorical perception may be regarded as the most elementary or analytical way of extracting information from a perceptual continuum.

The concept of categorical perception was originally developed to explain phoneme boundaries in speech sounds [Liberman et al. 1957, 1961]. Perceptual discrimination is normally easier across category boundaries. In other words, stimuli are more likely to be judged as "different" if they fall into different perceptual categories.

A familiar example of categorical perception is the perception of colour. Electromagnetic radiation in particular frequency bands evokes particular colours. The band of frequencies corresponding to a particular colour (red, orange, yellow, etc.) corresponds to a perceptual category.

The position of the boundary between two neighbouring perceptual categories is always somewhat vague or flexible. In a rainbow, for example, one cannot see exactly where "red" stops and "orange" begins. The position of the boundary between two categories also depends on the observer and on the context in which a stimulus is presented. For example, the colour aqua will sometimes be called blue, sometimes green, depending on observer and background colour.

The positions of category boundaries may be either innate or learned. Boundaries between colours appear to be primarily innate (due to the physiology of the eye). Boundaries between speech phonemes appear to be primarily learned by exposure to speech: adults' discrimination at phoneme boundaries is sharper than infants' [Eimas et al. 1971]. Similarly, the musical interval discrimination functions of musicians are sharper than those of untrained listeners, implying that boundaries between musical scale degrees are also learned [Burns and Ward 1978]. Innate forms of categorical perception are universal. For example, primary colour labels have similar or identical meanings in different languages. Learned forms, such as the categorical perception of speech vowels and musical intervals, are culture-specific.

The width of a perceptual category generally exceeds one difference threshold (or just noticeable difference, or difference limen). For example, optical frequencies which can be distinguished in only 50% of experimental trials may be regarded as one difference threshold apart; in ordinary perception, such frequencies normally fall in the same category, i.e. they have the same colour.

2.3.3 Holistic Perception and Pattern Recognition

Holistic (synthetic, global) perception is the perception of whole objects or scenes. It involves the direct extraction of high-level information from the environment. By contrast, analytic perception occurs only when a specific object or stimulus, or part thereof, is attended to. How holistically or analytically an even will be perceived depends on the observer [see e.g. Zenatti 1985] and on the context of the event.

Both percepts and sensations may be either holistic or analytic. An analytic sensation is defined to be the experience accompanying the "sensing", with an analytic attitude, of a stimulus. Holistic sensations are generally more meaningful than analytic sensations. They are also more likely to be linked to environmental objects, in which case they become "holistic percepts".

Holistic perception normally occurs quite spontaneously, with little or no apparent effort on the part of the observer. This is readily explained in the direct perceptual approach of Gibson [1966], according to which holistic sensations are merely experiences accompanying the direct perception of whole objects. Analytic perception requires an "analytic attitude", and can be quite difficult, even though the information being sought is more closely related to the information output by the sense organs than that sought in holistic perception. For example, it is quite difficult to hear out the harmonics of a complex tone.

Traditional psychophysics tends to regard analytic sensations as more fundamental than holistic sensations. This is because psychophysics is concerned with the relationship between sensations and the stimuli (such as light and sound patterns) which evoke them. This relationship is held to be mediated first by the physical-physiological transducing properties of the sense organs, and secondly by *perceptual grouping* processes, by which analytic sensations corresponding to physiological output of the sense organs are grouped by stages into holistic sensations.

General principles of perceptual grouping were described by Wertheimer [1923] and Koffka [1935]. The principles cover the grouping of both simultaneous and sequential events in music, i.e. both chords and melodies. Applications in hearing and music have been described in detail by Deutsch [1982b] and Moore [1982].

If the same sensation occurs at different times, the two events may be perceived to be related (and therefore to be likely candidates for perceptual grouping) due to their *identity*. Different stimuli are perceived as identical if their difference is not perceived, i.e. if they are close enough to be assigned to the same perceptual *category* (Sect. 2.3.2).

Sensations in different categories may be grouped by *proximity* if they are close on some psychophysical scale. Visual sensations are grouped if corresponding regions of excitation are nearby on the retina. For example, a dotted or broken line is perceived as such because the dots or line segments making it up are close to each other. Stars which in three dimensions are relatively far from each other are nevertheless perceived as constellations because corresponding points on the retina are close to each other. Spontaneous grouping of auditory sensations by proximity is called *streaming* (Sect. 2.4.6).

Grouping of sensations by *familiarity* is called *pattern recognition*. Familiar patterns of sensations correspond to regularitics or invariances in the environment [Gibson 1966; Bregman 1981]. Pattern recognition normally occurs quite spontaneously, with no conscious effort by the observer. The recognition of familiar patterns is an essential ingredient in the interaction of an organism with its various environments.

Instinctive behaviour in animals and humans is evidence that some aspects of pattern recognition are innate. However, most perceptual patterns become familiar by spontaneous learning and exploration in early life, implying that most aspects of pattern recognition are acquired. Later in life, pattern recognition processes become increasingly resistant to change: new perceptual patterns become increasingly difficult to learn and recognize.

Patterns may be recognized if they are incomplete, or if extra components are included. For example, a written word may still be recognized if some letters are added or taken away (i.e. if it is misspelled); the more letters are added or deleted, the less likely it is that the original word will be recognized. Melodies may be recognized if appropriate pitches are heard at appropriate times, in spite of missing or added notes: a melody whose notes are interleaved with distractor notes can still be recognized [Dowling et al. 1987].

The recognition of incomplete or superposed patterns may be modelled by *template matching* [cf. Uhr 1963]. A template (or prototype) is an idealized representation of the perceptually relevant features of a familiar pattern of sensations. Pattern recognition may be regarded as a process whereby matches are sought between the components of a template and configurations of sensations occurring in real time. The more components of the real-time configuration match those of the template, the more likely it is that the corresponding pattern will be recognized. Note that pattern recognition templates exist only as parts of perceptual models; they have no actual physiological correlates in the peripheral or central nervous system.

The classification of perceptual grouping criteria into identity, proximity and familiarity is not always clear cut. Familiar patterns are identical to or close to previously experienced patterns, and there is no sharp dividing line between identical and proximate sensations, due to the flexibility of perceptual categories.

2.3.4 Ambiguity, Multiplicity and Context

A stimulus is *ambiguous* if it may be interpreted in two or more different ways. Consider again the example of a misspelled word. The more letters are added or taken away from the original word, the more ambiguous the interpretation of the word becomes – unless, of course, a new word is formed with a new, unambiguous meaning. In the template approach to pattern recognition, a stimulus pattern is ambiguous if it may be matched by a number of different templates, or by the same template in a number of different ways.

Perceptual ambiguity is normally associated with holistic perception, in which a perceptual event can have only one meaning at a time. In analytical perception, an event is analyzed into a number of simultaneous percepts: the event exhibits perceptual *multiplicity*. In the case of a written word, for example, the reader's attention can switch from holistic to analytical perception, resulting in awareness of individual letters.

The same stimulus may be ambiguous or multiple or both, depending on its *context*. A single word on a blank page (e.g. "can") is ambiguous – it has several possible meanings. It is also multiple in the sense that one's attention is focused on the individual letters of the word. In context (e.g. "I drank a can of beer after work"), both ambiguity and multiplicity are reduced. Similarly, the pitch of a single complex tone in isolation may correspond to the pitch of its first or second harmonic (or both, or a number of other possibilities: see Sect. 6.1.5), but in the context of a melody the pitch rarely differs from that of the fundamental.

By reducing ambiguity, context facilitates comprehension. Letters are easier to read in words than they are in isolation, and words are easier to read in grammatical than in non-grammatical phrases [Cattell 1886]. Similarly, musical notes are easier to read in more "grammatical" tonal contexts [Sloboda 1976].

Ambiguity is relatively unusual in perception and language. In ordinary settings, perceptual patterns are overspecified: much of the information in the patterns is redundant [Garner 1970]. In natural language, ambiguity is normally avoided, for obvious reasons. In music, however, ambiguity plays an important role, maintaining interest and generating multiple expectations [Meyer 1973; Thomson 1983]. From this point of view, it is inappropriate to describe music as a language. If music *is* a language, it is more similar to poetry than to prose.

2.4 Tone Sensation

2.4.1 Terminology

Pipping [1895] distinguished between two kinds of pitch: the pitch of an individual pure tone component, such as a harmonic of a complex tone (which he called "tone pitch") and the overall pitch of a complex tone, corresponding to the fundamental frequency ("clang pitch"). Pipping thought that clang pitch was due to nonlinear distortion in the form of difference tones. Schouten [1940] observed that a complex tone appears to have two sensory components at the pitch of the fundamental, "one of which, having a pure tone-quality is identical with the fundamental tone, whereas the other, having a sharp tone quality and great loudness, is of different origin" (p. 358). Schouten called this additional subjective component the *residue*, hypothesizing that its pitch corresponded to the periodicity of upper, unresolved components of the complex tone.

Terhardt [1972, 1974a] made the same distinction as Pipping and Schouten, but used different terms and a different explanation for the two kinds of pitch. He proposed that *virtual* (clang/residue) pitch was formed by the spontaneous recognition of the familiar pattern of *spectral* (tone) pitches of a complex tone. The term virtual pitch added to a whole array of names for clang/residue pitch which had come into use in the meantime, among them fundamental pitch, periodicity pitch, and low pitch.

According to the American Standards Association [1960], there is only one kind of pitch: "Pitch is that attribute of auditory sensation in terms of which sounds may be ordered on a scale extending from low to high". This definition states that pitch is an *attribute* of auditory sensation – not a sensation in itself. The definition implies that there may be different kinds of *sensation* which *have* pitch, but there is only one kind of pitch. It is therefore appropriate in the above discussion to refer to two kinds of tone sensation rather than two kinds of pitch.

For this purpose, I have coined the terms *pure tone sensation* and *complex tone sensation*, as they refer directly to the types of tone which normally produce the two kinds of tone sensation. Spectral pitch may be defined using this terminology as the pitch of a pure tone sensation; virtual pitch, as the pitch of a complex tone sensation.

Pure and complex tone sensations, like all tone sensations, also have the attributes timbre and salience. This fact is hard to express using the terms spectral and virtual pitch. To refer to the timbre or salience of a spectral or virtual pitch is to refer to an attribute of an attribute. It is more logical to speak instead of the timbre or the salience of a (pure or complex) tone sensation.

A complex tone, in the proposed terminology, may evoke several different pure tone sensations (corresponding to its audible harmonics) and several different complex tone sensations (corresponding to implied fundamentals of different groups of harmonics). However, a (pure or complex) tone sensation is in all cases a single entity in the experience of the listener, with just one pitch, one timbre and one salience.

2.4.2 Pure Tone Sensations

Pure tone sensations are single sensations normally evoked by pure tones or pure tone components. They may also be produced by noise, because noise can evoke pitch [Fastl 1971]. Narrower bands of noise are more tone-like or "tonal" than wider bands [Aures 1984].

Complex tones are overwhelmingly heard as single wholes. The hearing out of pure tone components requires an unusually analytical listening attitude. Consequently, most people are unaware that this is possible. As hearing out of pure tone components is rarely necessary in musical performance, even musicians do not always develop the skill. For example, Rameau developed his theory of the *basse fondamentale* by experimenting with a Pythagorean monochord, and only afterwards learned that the harmonics of a tone could be individually heard [Christensen 1987].

Interestingly, Rameau believed that octave multiples of the fundamental frequency (the second, fourth, eighth, . . . harmonics) were inaudible in ordinary complex tones. Scientists such as Helmholtz and Stumpf agreed that the second and fourth harmonics were harder to hear out than the third and fifth [Plomp 1964]. This effect has not (to my knowledge) been backed up by experimental data. Perhaps it is due to musical conditioning, via octave equivalence. In any case, the effect is neither expected nor explained on the basis of Terhardt's [1972, 1974a] pitch theory.

The configuration of pure tone sensations in a sound may be represented by a graph of salience (perceptual importance) against time, called the *spectral pitch pattern*. This may be regarded as the ultimate basis for the sensory attributes (pitch, timbre, salience, etc.) of complex tone sensations [Stoll 1982]. The spectral pitch pattern may be modelled as a continuous function of time by Fourier time transform [Terhardt 1985; Heinbach 1986]. The recognition of patterns (and hence sound sources) among the contours of the spectral pitches in the pattern is remarkably analogous to the recognition of visual objects from the contours of their edges and boundaries [Terhardt 1986; cf. Gibson 1979].

The salience of a pure tone component (and hence of a pure tone sensation) may be defined as its probability of being noticed, or the degree to which it contributes to the perception of complex tones (Sect. 4.4.2). It depends on audibility (level above masked threshold) and, to a lesser extent, on frequency (Sect. 6.1.2). It also depends on context. For example, pure tone components are easier to hear, and therefore more salient, if they move relative to each other [Brink 1982]: this indicates that they do not come from the same source (e.g. they are unlikely to be harmonics of the same fundamental [McAdams 1984]).

The exact pitches of pure tone components within a complex sound depend not only on frequency but also on level and masking [Terhardt 1972, 1979a; Hesse 1987] (Sect. 3.3.3). Variations of pitch with masking and changes of level are called *pitch shifts.* The pitch of a low-frequency pure tone falls slightly as its level is increased [Stevens 1935]. For example, the pitch of the electric bass of a rock band can sound sharp relative to the rest of the music when the music is damped to barely audible level by walls and/or distance. Two simultaneous pure tones which partially mask each other have the effect of pushing each other apart in pitch by a small (but perceptible) amount [Walliser 1969c]. The effect of masking on pitch is seldom noticeable, as the pure tones concerned are normally perceived as components of complex tones, and the pitch of a complex tone is affected relatively little by masking [Stoll 1985].

2.4.3 Complex Tone Sensations

Complex tone sensations are generally associated with percepts (or "auditory images" [McAdams 1984]) of complex tones, such as people talking, and musical instruments being played. Most tone sensations in music and in everyday sounds are of the complex kind.

With respect to pure tone sensations, complex tone sensations are holistic: they are associated with the grouping or "fusion" ("Verschmelzung" [Stumpf 1898]) of pure tone sensations. Complex tone sensations may themselves combine to form other sensations such as chord and melody sensations in music. With respect to such higher order sensations, complex tone sensations are analytical.

A complex tone may be perceived as a whole even if its fundamental is missing, i.e. if it is a *residue tone.* Schouten [1940] theorized that the pitch of a residue tone depends on the *periodicity* of unresolvable higher frequency components – the "residue". This inspired decades of psychoacoustical research into the detection of periodicity among spectral components of complex tones [Moore 1982 and references therein]. Periodicity was supposed to be detected in time intervals between peaks in the fine structure of the waveform of a sound, and coded as synchronies (*phase-locking*) in neural firing patterns.

The periodicity model explains the spectral pitch of low-frequency pure tones, and the residue pitch produced by (apparently) unresolvable high harmonics. However, the model has some serious drawbacks. The underlying assumption that a direct correspondence exists between experience and brain states or processes is unscientific, or at best premature (Sect. 2.1.2). No physiological or anatomical evidence has been found for an appropriate time measuring mechanism. And as yet no one has been able to establish a model based on periodicity which makes sensible predictions concerning the pitch properties of complex sounds in the general way that the model of Terhardt et al. [1982b] does (although plans for such a model are described by Moore [1982]).

According to Terhardt [1972, 1974a] and others [Goldstein 1973; Wightman 1973], the (virtual) pitch of a complex tone results from the recognition of a harmonic pattern among the (spectral) pitches of its resolvable (i.e. audible) pure tone components. Terhardt's model differs from the others in that it is based on *familiarity* with the pitch pattern produced by ordinary complex tones [cf. Whitfield 1967]. In Terhardt's version of pitch pattern recognition, the physiology of the perception of pure tones – in particular, whether their pitch is determined by place or time information on the basilar membrane – is not relevant (Sect. 2.2.2). The basic data of the model are in no sense the "temporal patterns of firing in different groups of auditory neurones", as suggested by Moore [1982, p. 127]. The pattern recognition part of the model is concerned with the functional relationship between two sets of *experiential*, not physical parameters: the (spectral) pitches and audibilities of the pure tone components of a sound, and the (virtual) pitches and saliences of the complex tone sensations it evokes.

Why does the pitch of a complex tone correspond to the lowest component of the pattern (the fundamental) rather than some other component? A possible reason is that the pitch of the fundamental corresponds to the period of the complex tone's waveform [Rasch and Plomp 1982]. According to the pattern recognition model, however, the auditory system is not sensitive to the period of the waveform as a whole; temporal patterns are reflected by the *roughness* of a tone, not its pitch [Terhardt 1969]. Another possible reason is that the fundamental is normally the most audible (or salient) of the harmonics of a typical complex tone: it is only masked from one side, and from a considerable pitch distance (an octave), whereas the other harmonics are masked from both sides, and at smaller intervals [Terhardt 1979a]. (Note that the fundamental does not necessarily have the highest sound pressure level, SPL. Often, a higher harmonic has the highest SPL, e.g. if it falls in the centre of a speech vowel formant.) However, the audibility of the fundamental differs from that of the other components only by degree. Perhaps the unique property which distinguishes the lowest component from the others is simply that it is the lowest [Terhardt, personal communication]. The harmonic number of the highest audible component of a typical complex tone varies over a wide range – say, from about 5 to 15 (see Sect. 6.1.1) – but the harmonic number of the lowest audible component is almost always one.

The recognition of harmonic patterns among spectral pitches may be modelled by means of a *harmonic template* incorporating the salient features of the spectral pitch pattern of a typical complex tone (Sects. 2.3.3, 4.4.1) [cf. Cohen 1984]. The pitch distances between the components of the template are slightly stretched relative to a harmonic series of frequencies, due to pitch shifts [Terhardt 1979a]. The dependence of virtual pitches on spectral pitches [Houtsma and Rossing 1987] may be modelled by shifting the template across the pitch range and looking for matches between template components and real-time spectral pitches. The pitch of modelled complex tone sensations corresponds to that of the lowest template component; their salience, to how many spectral pitches match template components and how closely they match. Salience depends also on the context of other (pure and complex) tone sensations. Optimal fit is more important in the *spectral pitch dominance region* between about 300 and 2000 Hz [Terhardt et al. 1982b] (Sect. 6.1.2) than in higher or lower regions. So the virtual pitch of a complex tone does not necessarily correspond exactly to the spectral pitch of its fundamental, especially if the spectrum of the tone is slightly inharmonic. The template approach may be used to explain why and how, and to estimate to what extent, complex tones exhibit pitch shifts [Stoll 1984; Terhardt and Grubert 1987].

The recognition of harmonic pitch patterns in ordinary complex tones is universal. A remarkably analogous *cultural* aspect of tone perception is the assignment of tones in a musical context to particular steps of a diatonic scale, i.e. the recognition of *diatonic* pitch patterns. Jordan and Shepard [1987] studied this by presenting listeners with major scales whose intervals had been uniformly stretched (so that the octave was noticeably larger than normal) or equalized (to produce 7-tone equal temperament). The resultant shifts in the pitches of other scale steps (notably the tonic) could be explained by postulating a rigid *diatonic template* (or tonal schema) consisting of scale steps separated by the familiar intervals of the major scale.

Harmonic and diatonic pitch pattern recognition are similar in the following ways. Features of the pattern-recognition template are acquired by experience of regularly recurring pitch patterns; pitch intervals between template elements remain the same in spite of irregularities in input stimuli; modelling involves finding the *best fit* between the template and some configuration of pitches heard in real time [Moore et al. 1985]; pitch ambiguity effects (of both complex tone sensations and tonics) may be explained in terms of alternative template fits; and pitch shift effects (again, of both complex tone sensations and tonics) may be accounted for in terms of the lining up of template components. In both cases, it should be emphasized that the template is no more than part of a model and has no physiological reality.

The musical pitch of tone sensations becomes difficult to judge above a frequency of 4–5 kHz [Bachem 1948]. Proponents of the periodicity approach to pitch perception believe this is due to uncertainty in the time at which nerve impulses begin, which prevents phase-locking above 4–5 kHz [Rose et al. 1967]. Proponents of the pattern-recognition approach point out that speech harmonics rarely have audible harmonics above 4–5 kHz (e.g. the eighth harmonic of 500 Hz) and so the auditory system is not familiar with harmonic pitch patterns in this region [Terhardt 1979a]. Whatever the reason, pure tone sensations above 4–5 kHz (i.e. above the top end of the modern piano) practically never play a harmonic role in music. The tones of the top two octaves of the piano are normally heard not as complex but as pure tone sensations, corresponding to their fundamental pure tone components; the second and higher harmonics of these tones contribute to timbre, but not to pitch (Sect. 6.1.2).

A clear complex tone sensation may be evoked by three successive harmonics not including the fundamental [Fletcher 1924]. The complex tone sensation evoked by two such harmonics is weaker, but can still be heard under suitable conditions [Sutton and Williams 1969; Smoorenburg 1970; Houtsma 1979]. There is even evidence for the existence of subharmonic complex tone sensations of single pure tone sensations [Houtgast 1976] (Sect. 2.4.5).

2.4.4 Pitch Ambiguity of Complex Tones

The frequencies of the lower harmonics of a complex tone do not automatically specify the tone's fundamental frequency. For example, the second and fourth harmonics could be the first and second harmonics of a complex tone an octave higher, or the first and second harmonics could be the second and fourth harmonics of a (partially masked) tone an octave lower. This explains why the virtual pitch of a complex tone in isolation is *ambiguous* (Sect. 2.3.4): it may lie an octave or (occasionally) some other consonant interval (e.g. a fifth) above or below the main pitch [Terhardt 1972, 1974a; Terhardt et al. 1986].

The degree of ambiguity of the pitch of a complex tone – the likelihood of hearing a virtual pitch which doesn't correspond to the fundamental – depends on the spectrum of the tone, the presence of maskers, and the context in which the tone is heard. The probability that the second harmonic will be heard to be the fundamental increases as the relative audibilities of the upper harmonics increase. The influence of spectral envelope (timbre) on the octave position of the pitch of a complex tone has been demonstrated in musical interval recognition experiments [Hesse 1982]. Complex tones in the presence of maskers evoke fewer pure tone sensations than unmasked tones, and therefore specify the position of the fundamental less clearly, especially if the fundamental is absent [Terhardt et al. 1986]. Tones in context (e.g. in a melody) are less ambiguous with respect to pitch than isolated tones: context normally suggests which pitch register a tone is most likely to be heard in.

The pitch ambiguity of complex tones has several important consequences for music. Inexperienced performers sometimes try to tune their instruments to a frequency an octave away from that of a reference tone such as a piano tone; once they realize their mistake, they find it easier to attend to the "correct" (main) pitch of their instrument and of the reference tone. Children and people with little musical training or ability sometimes sing a fourth, fifth of octave away from the correct pitch of a melody. The first and fifth scale degrees of a melody are sung more faithfully than other scale degrees by young children [Francès 1972]. Sequential tones at intervals such as octaves, fifths and fourths are perceived to be related (Sects. 5.4–6), and notes an octave apart are given the same (or similar) names in written music (Sect. 3.3.1).

2.4.5 Subharmonic Pitches of Pure Tones

Pure tones seldom occur in isolation in the everyday environment, but sometimes only one of the pure tone components of a complex sound is audible, due to masking by other sounds. For example, loud background noises may make speech almost inaudible, so that occasionally only one of the harmonics of a speech vowel can be heard. Such a vowel may still evoke a complex tone sensation, whose (virtual) pitch lies in the range suggested by the context in which the vowel is heard, i.e. the range of pitch of previous speech vowels, or the range of pitch characteristic of a particular speaker. The *exact* pitch of the vowel may be determined by lining up the (spectral) pitch of the audible harmonic with one of the upper components of the harmonic template described in Sect. 2.4.3: it is a "sensory subharmonic" of the (spectral) pitch of the audible harmonic. A pure tone can thus be perceived either in the ordinary way, as a pure tone sensation, or – in certain contexts – as a complex tone sensation at a subharmonic pitch.

Experiments on the perception of melodies of pure tones in which selected tones are displaced by octaves [Deutsch 1973; Kallman and Massaro 1979] suggest that pure tones an octave apart can have similar melodic functions. Since pure tones appear in musical contexts in these experiments, it is likely that musical experience strongly influences the results, so it is not clear how much the results might be influenced by subharmonic pitches.

Similarity ratings of sequential pure tones by nonmusicians [Allen 1967; Thurlow and Erchul 1977] (Sects. 5.5, 6) suggest that pure tones an octave apart sound more similar than pure tones separated by slightly larger or smaller intervals. This effect could be due either to the lining up of the main pitch of one tone with the suboctave pitch of another (pitch commonality), or (again) to musical experience. By contrast, Kallman [1982] found no effect at all at the octave when pure tones were compared for similarity.

Experiments on the pitch of pure tones [Houtgast 1976] have demonstrated the existence of subharmonic pitches in the presence of background noise. Noise may allow non-existent harmonics to be spontaneously imagined [Brink 1982], and thence to contribute to the formation of complex tone sensations. In the absence of noise, Houtgast found no evidence for the existence of subharmonic pitches.

Demany and Armand [1984] familiarized infants aged three months with short melodic fragments of pure tones, and then presented new fragments in which selected tones had been shifted away from their original frequencies. The infants displayed fewer novelty reactions when tones were shifted through octave intervals than they did for other intervals. The authors concluded that pure tones an octave apart sounded similar to these infants. The results of Demany and Armand may be explained by the hypothesis that the subharmonic tone sensations of pure tones are more salient for infants than they are for adults, due to infants' lack of familiarity with isolated pure tones. Infants have extensive experience of complex tones (e.g. speech vowels), and so are sensitive to simultaneous patterns of pure tone sensations, but they normally have relatively limited experience of sound sources which are capable of producing pure tones, and hence evoking *single* pure tone sensations (e.g. loudspeakers in psychoacoustical laboratories).

In experiments on *octave stretch* [Ward 1954; Terhardt 1970], listeners are required to tune the octave interval between alternating high and low pure tones by adjusting the frequency of one of the tones. The frequency ratio between the tones is generally found to exceed 2:1 by a small but significant margin. Terhardt [1972, 1974a] explained this result as follows: each pure tone evokes a single, unambiguous pitch, and the listener compares the interval between these pitches with the interval normally heard between the lower two pure tone sensations evoked by a complex tone, an interval which is slightly stretched due to pitch shifts. An alternative (and, for practical purposes, equivalent) explanation is that the two tones are perceived to be similar due to their pitch commonality (Sect. 3.2.3): the listener lines up the suboctave pitch of the higher tone with the main pitch of the lower tone.

2.4.6 Melodic Streaming

A complex tone sensation may be regarded as a grouping of pure tone sensations resulting from the spontaneous recognition of a familiar, harmonic pattern. A melodic stream is another kind of perceptual grouping, either of pure or of complex tone sensations, due to *proximity* in one or more tonal attributes (loudness, pitch, timbre, duration) or in time.

Streaming of pure tone sensations due to proximity in pitch and time occurs both for adults [Miller and Heise 1950; Noorden 1975] and for infants [Demany 1982]. A directly analogous effect occurs in vision: a pair of lights switched on and off in alternation in a dark room look like a single, moving light, provided they are close enough and they alternate fast enough [Kubovy 1981].

Complex tones, when perceived as wholes, may stream if they are similar in *timbre* [Bregman and Pinker 1978; Wessel 1979]. In orchestration, woodwind parts blend better if they are "dovetailed" (e.g. if one oboe plays higher than the first clarinet and one lower) as this inhibits streaming by timbre. In ambiguous cases, a tradeoff occurs between streaming of pure tone sensations by pitch and of complex tone sensations by timbre, depending on the relative saliences of the tone sensations [Singh 1987].

Two or three auditory streams may be heard simultaneously, but it is difficult to attend to more than one [Bregman and Campbell 1971]. It is difficult to separate streams that cross over in pitch: interleaved melodies, in which the tones of two different melodies alternate, merge into a single, unrecognizable sequence if the melodies overlap in pitch [Dowling 1973]. However, the interleaved melodies are easier to recognize if they are already familiar.

Streaming is affected by timing and source direction. Simultaneous sounds are easier to discriminate (i.e. to segregate into different streams) if their onset times are not quite the same [Rasch 1978], as is usually the case in musical performance [Rasch 1979]. Sounds from similar directions stream (e.g. the "cocktail party effect", stereo reproduction of orchestral music). Perception of the direction of a sound source is assisted by head movements and vision [Lippman 1963b; Gibson 1966].

Sound sources (e.g. musical instruments against an orchestral texture) may often be identified by coherent variation of physical characteristics such as amplitude and frequency of harmonics [McAdams 1984; Bregman et al. 1985] otherwise known as *vibrato*. Vibrato makes it easier to follow a particular voice against to contrapuntal background. In Romantic opera, for example, vibrato enables solo voices to penetrate loud (or thick) orchestral textures. However, vibrato also inhibits blending of voices. This may explain why less vibrato was used in Baroque opera, where harmonizing was more important, and the music less passionate [Galliver 1969]. The blending of vibrato voices is improved if the vibrato is synchronized (e.g. in string quartets).

Like complex tone perception, melodic streaming may be regarded as consequence of *familiarity* with the auditory environment. In general, sounds stream if they appear to come from the same source [Bregman 1981]. Such sounds are often close in tonal attributes (loudness, pitch, timbre) and in time and direction, but not always. For example, the timbre of the clarinet differs markedly between its registers, but this does not necessarily inhibit the streaming of clarinet tones in music.

Sounds from different sources can stream, if it appears that they could have originated from the same source. In musical *hocket* the notes of a melody are played alternately on different instruments or sung by different voices. Because the timbre of the instruments or voices varies relatively little, the result sounds like a single melody. Examples are to be found in some African and Indonesian musics and, in the West, in medieval music (including Gregorian chant) and among the compositions of Webern [Dalglish 1978; Erickson 1982].

When two tones of different pitch and loudness alternate in *legato* (i.e. with no silent gap between them), the quiet tone may be perceived to remain sounding through the loud tone, even though it is physically absent [Thurlow and Elfner 1959]. This is an example of the effect called *closure* by the Gestalt psychologists. In this example, closure occurs only if the louder tone would have completely masked the quieter tone had the quieter tone actually been present. Intelligibility of speech in a noisy environment is enhanced by the effect of closure [Miller and Licklider 1950]. Like other streaming effects, this effect arises from familiarization with the audible world. It differs from other streaming effects in that it is determined not solely by regularities of the auditory environment but to a large extent by physiological limitations of the ear (i.e. masking). In other words, it is determined by the nature of the interaction between the organism and its environment.

After considering the available experimental evidence on streaming, Sloboda [1985, p. 162] concluded that "pitch streaming is a real 'pre-musical' phenomenon, although musical knowledge may interact with and modify its effects." Streaming may thus be regarded as a sensory basis for melodic perception, and so for the theory of counterpoint [Wright 1986]. Just as melodic streaming occurs when sounds are somehow close in their sensory attributes, melodic continuity and unity are enhanced in musical performance by maintaining a relatively constant dynamic (loudness); and in composition, melodic continuity and unity are maintained by the use of small pitch and time intervals, and by maintaining a particular orchestration. In mainstream music theory and practice, wide leaps are avoided in melodies and in the voices making up a harmonic progression; when wide leaps do occur, their disruptive effect is reduced by resolving the second note by stepwise movement in the direction of the first note.

2.5 Pitch Perception

2.5.1 Dimensionality

The generally accepted definition of pitch (Sect. 2.4.1) implies that it is a one-dimensional sensory continuum. The psychological reality of such a continuum is apparent from such elementary perceptual skills as the ability to identify the higher of two pure tones (an ability which is shared by infants [Trehub 1987]) and the ability to estimate the magnitude of the pitch distance between two pure tones [Stevens et al. 1937].

Musical pitch may be described as *multidimensional*, its two main dimensions being pitch height (as in the one-dimensional model) and tone chroma [Shepard 1964, 1982; Idson and Massaro 1978]. This may be concluded from multidimensional scaling solutions of experimental results on the similarity or relative height of complex tones (Sect. 1.3.2). However, such experimental results depend on the spectra of the tones presented to listeners [Ueda and Ohgushi 1987]. It would seem more straightforward to describe the pitch of complex tones (including octave-spaced tones) as *ambiguous* relative to a *one*-dimensional pitch continuum [Terhardt 1972, 1974a, 1979b]. This clarifies the distinction between sensory and cultural effects in musical pitch perception, especially the perception of octave equivalence (cf. Sect. 3.3.1; Chap. 5), and makes it unnecessary to postulate the existence of "cognitive structures" in order to account for experimental results.

2.5.2 Continuous Pitch Scales

Fletcher [1934] proposed measuring the pitch of a sound in terms of the frequency of a pure reference tone of constant loudness, whose pitch is judged to be the same as that of the sound. This measure of pitch, which may be called *equivalent frequency*, was also used by Terhardt [1972, 1974a]. Terhardt's method was identical to that of Fletcher, except that he held the sound pressure level of the pure tone constant instead of its loudness. This procedure is analogous to that for measuring *loudness level* (Sect. 2.2.1), in which the frequency of a pure reference tone is held constant (at 1 kHz) and its SPL is varied; loudness level could be called "equivalent SPL".

Equivalent frequency and equivalent SPL are not proportional scales: doubling equivalent frequency does not necessarily make a sound seem twice as high, nor does doubling equivalent SPL make a sound seem twice as loud. Stevens et al. [1937] developed a proportional pitch scale (analogous to *loudness* in sone) called the *mel scale*, in which equal scalar intervals (measured in *mel*) corresponded to equal apparent interval sizes [see also Stevens and Volkmann 1940; Beck and Shaw 1963]. The mel scale is roughly proportional to the logarithm of frequency above about 1 kHz, and approaches a linear relationship with frequency at low frequencies [Zwicker and Feldtkeller 1967]. Like loudness in sone (Sect. 2.2.1), pitch in mel is quite imprecise; it is unsuitable for measuring small pitch effects such as pitch shifts.

The mel scale is an appropriate measure of pitch only when *pure* tones are heard in a *non-musical* context by musically untrained listeners. For example, Attneave and Olson [1971] found the scale to be inappropriate for the musical task of melodic transposition. Moreover, the mel scale, as originally defined by Stevens et al. [1937], only applies to the apparent size of intervals between pure tones of *equal loudness*, not equal SPL as in the experiments of Elmasian and Birnbaum [1984].

Pitch in mel may be scaled by the rule that equal sensory intervals contain roughly equal numbers of difference thresholds [Terhardt 1968c]. The difference threshold of frequency depends considerably on the listener, both for pure tones [Fastl and Hesse 1984] and complex tones [J. Meyer 1979]. On average, the difference threshold of frequency for sequential pure tones is around 0.05 semitones in the region above 500 Hz (cf. 1.0–1.5 semitones for simultaneous pure tones: Sect. 2.2.2). At lower frequencies, it is about 1 Hz [Fastl and Hesse 1984].

The difference threshold of frequency for complex tones is about the same as that for pure tones, with the exception that it retains its lowest (high-frequency) value (about 0.05 semitones) right down to about 100 Hz, or G_2, the bottom line of the bass clef [Walliser 1969b]. This may be understood in terms of spectral pitch dominance. The pitch of a complex tone of fundamental frequency 100–400 Hz normally depends only on the pitches of the 3rd, 4th and 5th harmonics [Ritsma 1967]. Complex tones with fundamental frequencies lower than about 100 Hz no longer have dominant harmonics above 500 Hz.

So the pitch difference threshold for a complex tone, measured in semitones, increases as its frequency falls below 100 Hz, i.e. as the frequencies of its dominant harmonics fall below 500 Hz.

Above 100 Hz, the apparent size of melodic intervals is proportional to their size in semitones. The log frequency or *frequency level* scale is therefore appropriate for the pitch of complex tones across almost all of the musical range. Only when melodies are transposed into the deep bass (below the bass clef) do melodic intervals sound smaller than normal.

2.5.3 Categorical Pitch Perception

The diatonic scales (major and minor) are familiar to members of Western culture, musicians and non-musicians alike. Also familiar are the non-scale notes which occur in diatonic music. In other words, the entire chromatic scale is familiar, provided a certain subset of that scale (the diatonic scale) is emphasized. Consequently, a pitch interval of random size is perceived by musicians to belong to a particular semitone category (m2, M2, m3, etc.) [Siegel and Siegel 1977a, b; Burns and Ward 1978]. Similarly, a complex tone of random frequency, presented in a tonal musical context, is perceived as belonging to a particular scale step.

The perceptual categorization of musical pitches and intervals may be regarded as a prerequisite for the understanding of pitch relationships and structures. Categorization reduces the amount of information carried by the pitches of a passage of music to a manageable level (Sect 2.3.2), removing information about the precise tuning of a pitch or interval, and retaining only its semitone category. Even under ideal listening conditions, mistunings of 0.1 – 0.3 semitones (depending on the interval) are acceptable [Moran and Pratt 1926; Vos 1982; Hall and Hess 1984]; even larger variations are acceptible in musical performances [Burns and Ward 1978, and references therein]. Perceptible out-of-tuneness does not necessarily affect musical meaning and function. For example, an out of tune subdominant chord still has a subdominant function within its key context, provided, of course, that it is not so out of tune that it is perceived as another chord. Mistuning is more disturbing for more salient pitches, e.g. those of a melody as opposed to its accompaniment [Rasch 1985].

The categorical perception of musical pitch begins when the auditory system "decides" whether a particular audible harmonic belongs to a complex tone [Moore et al. 1985]. Terhardt et al. [1982b] accounted for this decision-making process by assigning a "harmonicity" value to the interval between an audible component of a complex tone and its fundamental. In their model, calculated harmonicity falls gradually to zero when an interval is mistuned by 8% in frequency, or a little over a semitone. The harmonicity of the interval between a tone component and an assumed fundamental may be regarded as a measure of the probability that the component will be perceived as belonging to a complex tone with that fundamental.

In well-tuned Western harmonic progressions, the frequencies of audible pure tone components are close enough to equal temperament that all spectral pitches may be unambiguously assigned to degrees of the chromatic scale. The model in Chap. 4 takes advantage of this by defining all pitches and intervals (including intervals between harmonics and fundamentals) relative to the pitch categories of the chromatic scale. This simplifies the above decision-making procedure: the probability that a particular tone component is perceived as belonging to a particular complex tone in the model is effectively either 100% or zero. Categorization of pitch is also appropriate for modelling pitch commonality and pitch distance (Sects. 4.6.1, 2). Pitch commonality is concerned with sequential tone sensations in the same pitch category; pitch distance, with sequential tone sensations in different pitch categories.

2.5.4 Musical Training

As Western musicians belong to the class of Western audiences, all aspects of pitch perception discussed above apply for Western musicians. In addition, musicians' pitch perception is conditioned by their knowledge of the relationship between music as it sounds and as it is written and played, i.e. by their knowledge of music theory.

Only musicians have the opportunity to learn to recognize by name the notes, intervals, chords and keys of Western music from the sound alone. A prerequisite for the recognition of such musical elements is their categorical perception (Sect. 2.3.2).

Intervals are recognized primarily on the basis of their absolute sizes, not on the basis of their consonance. Evidence for this is that confusions between intervals in recognition experiments normally occur between intervals of similar size rather than similar consonance: for example, fifths are confused with sixths much more often than with octaves [Plomp et al. 1973; Terhardt et al. 1986]. Consonant intervals are nevertheless easier to recognize than dissonant ones [Terhardt et al. 1986]. This effect may be regarded as sensory (tones spanning consonant intervals have pitches in common, Sect. 3.2.3), cultural (consonant intervals occur more often in melodies [Jeffries 1974]), or indirectly sensory (consonant intervals occur more often in melodies because of their pitch commonality).

Differences between the perception of music by musicians and non-musicians are important for experimental reasons (Chap. 5). Musically trained participants in psychoacoustical experiments can respond in quite different ways from untrained listeners. They may be "better" participants in that their responses are more consistent and more sensitive to subtleties. If sensory properties of sounds are of paramount interest, as in this study, musicians may be "worse" participants in that their responses are more strongly influenced by musical conditioning.

2.5.5 Perfect Pitch

Everyone has *absolute pitch* in that they can discriminate male and female adult voices by their pitch alone (e.g. on the telephone). This kind of absolute pitch has an uncertainty or category-width of, say, three to six semitones. Like other aspects of absolute pitch, it is based on experience. Experiments with infants [Clarkson and Clifton 1985] suggest that the minimum uncertainty of "universal" absolute pitch is probably about three semitones. The fact that this is about the same as the width of a critical band in the most important range of pitch (Sect. 4.3.1) is probably coincidental: critical bandwidth is only important for simultaneous tones, whereas absolute pitch applies to isolated tones.

In Western music, the pitch continuum is divided up into absolute pitch categories much smaller than those of speech, corresponding to steps of the chromatic scale (Sect. 3.3.2). Normally, only the performer of a piece of music is aware of the names of these categories ("F", "A♭", etc.). Therefore, only musicians are in a position to develop that kind of absolute pitch, called *perfect pitch*, in which pitch is identified absolutely in semitone categories [Harris and Siegel 1975]. Perfect pitch is normally acquired in childhood [Sergeant 1969]. With sufficient practice, it can also be acquired later in life [Cuddy 1968; Brady 1970].

There are many theoretical approaches to the origins and nature of perfect pitch [e.g. Ward 1963; Ward and Burns 1982; Costall 1985; Heyde 1987]. They mostly concentrate on perfect pitch in Western music. However, Western music is more highly developed regarding *relative* pitch (i.e. harmony) than absolute pitch. It may therefore be fruitful to look at perfect pitch (i.e. absolute pitch identification with an accuracy of a semitone or less) in musical cultures where harmony is less important. For example, unaccompanied melodies in Australian aboriginal music are sung in different places and at different times at the same frequencies, with an uncertainty of less than a semitone [Ellis 1967].

A surprising thing about perfect pitch is that so few musicians develop the ability, considering that absolute identification of stimulus properties is normal in the other senses [Watt 1917]. The reason why so few Western musicians have perfect pitch may be due in part to a conflict between the *spontaneity* of absolute pitch judgments and the *analytical* attitude to pitch required of the Western musician. The verbalization of musical note names requires quite an analytical attitude, perhaps because there are so many notes to distinguish between [Miller 1956]. In spite of this, perfect pitch often occurs quite spontaneously, in the following ways. Folk and ethnic musics in which a kind of perfect pitch is in evidence tends to be performed in a more spontaneous manner than Western art music. Perfect pitch is aided by other, relatively spontaneous experiences such as strong emotive associations [Ellis 1985, p. 65], chromesthesia or "colour hearing" [Peacock 1984; Rogers 1987], and the realization that a familiar passage of music is being played in the right or the wrong key [Terhardt and Ward 1982; Terhardt and Seewann 1983].

Musicians who identify familiar musical tones (e.g. piano tones) with great reliability are not so good at identifying the pitches of pure tones [Lockhead and Byrd 1981], suggesting that *timbre* plays an important role in perfect pitch. Absolute timbre perception and absolute pitch perception are hard to separate; this may be regarded as an example of the general fuzziness of the distinction between pitch and timbre (Sect. 2.2.1).

In a "direct perception" approach [Gibson 1966], absolute pitch involves the identification of sound *sources* according to their pitch. This explains the spontaneity of absolute pitch judgments. Further, since sound sources which differ in pitch (e.g. different piano strings) also differ in timbre, it also explains why timbre sometimes interferes with absolute pitch judgments.

Terhardt's [1972, 1974a] approach to pitch perception yields new insights into perfect pitch. According to Terhardt, musical tones exhibit octave ambiguity (the octave position of the pitch of an isolated complex tone is somewhat uncertain) and pitch shifts (the pitch of a complex tone is slightly different from that of pure tone of the same frequency). As perfect pitch possessors "memorize" the sensory properties of musical tones, they inevitably "memorize" the tones' octave ambiguity and pitch shifts, and these properties inevitably affect absolute pitch judgments. This readily explains the octave and semitone errors found by Balzano [1984] in experiments on the absolute pitch of pure tones.

3. Psychomusicology

The perception of music, and hence its performance, composition and historical development, is conditioned by the repeated exposure of musicians and audiences to auditory patterns and regularities. Such patterns occur in music itself, in everyday social and physical environments, and before birth. Musical consonance depends on the smoothness and tonalness of tone simultaneities, the pitch commonality and pitch proximity of sequences, and cultural conditioning. Pitch in tonal music is specified relative to diatonic and chromatic scales, in which notes an octave apart are harmonically equivalent; intonation within chromatic pitch categories involves a compromise between minimum roughness and maximum pitch commonality. Diatonic (major/minor) tonality depends on various aspects of consonance and musical pitch, but especially on the roots of chords and broken chords.

3.1 Conditioning

3.1.1 Sensory Versus Cultural

Ever since Pythagoras's conflict with Aristoxenes [Cazden 1958], music theorists have been trying to separate "nature" from "nurture." For example, Rameau [1721] and Schenker [1906] invoked the supposed "naturalness" of simple frequency ratios and the harmonic series to explain harmony. Cazden [1954], at the start of his criticism of Hindemith [1940], divided contemporary discussion of the fundamental questions of music theory into "three main types: those claiming to be founded on the laws of Nature, those claiming the priority of the arbitrary instinct of the composer, and those which rely on the observed practice of the art of music ... the natural, the subjective and the empirical theories". More recently, DeWitt and Crowder [1987] distinguished between "rational" musical theories with an acoustical basis in "natural law" (simple integer ratios, harmonic series, etc.) and "empirical" theories based on context and learning.

In a psychoacoustical approach, the "natural" approach to the origin of harmony is rejected, as it fails to account for the difference between physical reality and the way it is perceived (Sect. 1.4.1). The nature/nurture distinction becomes a distinction between *innate* and *acquired.* Innate aspects of perception are supposedly dependent on the physiology of the sensory organs and the nervous system; acquired (or "learned", or "conditioned") aspects, on familiarity with regularities in the human environment. The innate/acquired distinction is similar to the distinction between "phylogenetic" (i.e. related to evolution) and "ontogenetic" (related to individual development).

The auditory system is conditioned by different patterns of sound at different developmental stages and in different situations. The foetus is repeatedly exposed to the internal sounds of the mother's body, of which the heartbeat, walking sounds and the voice appear to be the most relevant for music (Sect. 3.1.2). After birth, infants, children and adults are repeatedly exposed to sound patterns (in pitch and time) which indicate the presence of particular objects (sound sources) in the environment. In a particular musical culture, individuals are repeatedly exposed to the particular, arbitrary patterns of sound characteristic of the culture's musical styles. Thus, *music may be regarded as a multilayered structure of more or less familiar patterns, and its perception seen as a multilayered process of pattern recognition, based on familiarity.* In the words of Davidson et al. [1987, p. 606], "it would seem that, in the experiencing of music, familiarity is the most important variable."

From a musical viewpoint, the most important aspect of the ear's physiology is the spectral analysis of complex sounds into pure tone components enabled by the cochlea and basilar membrane (Sect. 2.2.2). The process of becoming familiar with and learning to recognize environmental sounds in early life is largely a process by which the central nervous system becomes attuned to auditory sound spectra as functions of time, output from the cochlea as patterns of neural activity.

The nature of these spectra depends as much on the physical nature of environmental sounds (e.g. the periodicity of the waveform of a tone; Sect. 1.2.6) as it depends on the physiology of the ear. In Gibson's [1966, 1979] approach, both of these belong to the same "environment"; both innate and acquired characteristics originate from a single integrated system. This is particularly true in the case of prenatal conditioning (Sect. 3.1.2).

In the present study, aspects of perception dependent on physiological (innate) properties and limitations and on experience of universal aspects of the human environment are brought together under the one umbrella and called *sensory*. Aspects of perception associated with consistent and systematic differences in human environments, especially social and musical environments, are called *cultural*.

The boundary between sensory and cultural, like the boundary between innate and acquired, is not cut and dried. Both sensory and cultural aspects of perception involve conditioning through perception of and interaction with physical and social environments. In general, it is hard to know where to draw the line between aspects of the perceptible world which are universal and aspects which are culture-specific, and hence between aspects of perception which are conditioned by one or the other of these. In the present study, the only universal environmental sound taken into consideration is the *harmonic complex tone*, which is fundamental to both speech and (Western) music. Experimental effects (Chap. 5) and aspects of music theory (Chap. 6) are described as "sensory" if they appear to be a result of exposure to complex tones.

Both sensory and cultural forces significantly influence the nature and development of music. Different musical styles are constrained by the same universal (sensory) properties of human hearing. These lead to cross-cultural similarities, such as the widespread use of the octave, fourth and fifth intervals in world musics, and the independent evolution of the chromatic scale in different musical cultures [Burns and Ward 1982]. On the other hand, cultural differences (different technologies, social conventions, attitudes to music, etc.) and the flexibility and adaptability of human perception combine to produce large intercultural differences in musical style. Thus, Westerners are conditioned to Western tonal styles [Francès 1972; Cuddy et al. 1979], and members of different cultures perceive the same music (or musical fragments) in different ways [Castellano et al. 1984; Kessler et al. 1984; Davidson et al. 1987].

The importance of both sensory and cultural effects in music perception is repeatedly acknowledged, although the "sensory" aspect is given different shades of meaning by different authors. Helmholtz [1863] recognized the important roles of both "physical and physiological acoustics" and "musicology and aesthetics" for the structure of scales and chords [Landau 1961]. Lerdahl and Jackendoff [1983] explicitly acknowledged the separate roles of sensory and cultural effects in the development of music-theoretical conventions by dividing their well-formedness and preference rules for the analysis of tonal music into those apparently reflecting innate perceptual or cognitive capacities and those apparently acquired by exposure to Western music. More recently, Roberts [1986] described experimental results which "support both psychoacoustic and relativistic theories of consonance and dissonance" (p. 170).

Cultural conditioning is often thought of as a kind of *brainwashing* – a process by which listeners' perception of music becomes biased toward the arbitrary structural characteristics of particular musical styles. Another aspect of cultural conditioning is *sensitization.* Repeated exposure to music makes listeners more sensitive to sensory aspects of the structure of particular kinds of sound, such as those modelled in Chap. 4 (see Sect. 5.8.2). For example, Westerners may be insensitive to the intricacies of African drum rhythms, while Africans may be insensitive to the intricacies of Western pitch structures. Aspects of African drum rhythms may nevertheless derive from universal aspects of pulse perception [Parncutt 1987b], and aspects of Western harmony may be based on universal aspects of the perception of complex tones [Terhardt 1974a, 1976].

3.1.2 Prenatal Conditioning

Perceptual learning (or conditioning) by repeated exposure to particular sound patterns occurs throughout life. Presumably, the process begins as soon as perceptual systems begin to function in the foetus. The process then accelerates until, at birth, it has gained enough momentum to enable the infant to cope with the suddenness and surprises of the postnatal world. An investigation of the first, prenatal stage of auditory conditioning suggests that it may have a significant influence on the perception of music by individuals, and on the historical development of music.

There exists considerable indirect evidence that both sensory and cultural aspects of musical conditioning begin before birth. In the case of cultural conditioning, there is anecdotal evidence from musicians whose mothers listened to or played specific pieces of music during pregnancy [Verny 1981]. The present study is concerned only with prenatal conditioning at the sensory (or universal) level: conditioning of the auditory system of the foetus by the internal sounds of the mother's body [Parncutt 1987a]. If the prenatal conditioning hypothesis is correct, these sounds influence auditory perception, and especially music perception, across cultures.

Recent experimental evidence [Trehub 1987 and references therein] has demonstrated that infants are remarkably sensitive to elementary musical structures. Specifically, they are sensitive to pitch contour and streaming (elements of melody), musical intervals such as the octave, fifth and third (elements of harmony), and long and short events in rhythmic patterns across tempo variations (elements of rhythm). This sensitivity is unlikely to be innate, as it doesn't directly improve the infant's chances of survival. Nor is it likely to be the result of musical conditioning – infants are presumably too busy learning how to interact with their environment (people, food, ...) to pay much attention to the musical sounds to which they may be exposed.

Prenatal conditioning provides a more straightforward and tangible explanation of infants' sensitivity to musical elements. Sensitivity to melody and harmony may be a simple and direct consequence of repeated prenatal exposure to the mother's voice; sensitivity to rhythm, a consequence of repeated prenatal exposure to the mother's footsteps and heartbeat.

The ability of the foetus to hear is well documented. Unborn sheep respond to sounds from outside the mother from 6–7 weeks before term [Bernard et al. 1959]. In humans, both the inner and outer ear are completely developed about twenty weeks before term; from this point on, the foetus responds to sounds [see e.g. Verny 1981].

The prenatal auditory environment is loud and varied [Armitage et al. 1980]. In sheep, Bench [1968] referred to internal background sounds of around 72 dB. Sounds audible to foetal sheep are associated with drinking, eating, breathing, vocalizations, digestion, heartbeat and movement of the mother; sometimes the foetal heart is also audible to the foetus [Vince et al. 1982b]. Presumably, the human foetus is exposed to a similar range of sounds. Those associated with breathing, vocalizations (speech), heartbeat and movement of the mother include regularly repeating, and therefore easily recognizable, patterns (as further described below). Sounds associated with drinking, eating and digestion appear to be too irregular to influence mainstream Western music in obvious ways (although they could, perhaps, be exploited in electronic or non-Western musics).

Prenatal auditory stimulation affects postnatal behaviour in measurable ways. In the case of sheep, prenatally familiar sounds can cause sigh-like changes in respiratory patterns; they also produce less heartbeat acceleration than unfamiliar sounds [Vince et al. 1982a]. Sounds heard by birds before hatching affect their posthatching behavior [Vince 1980]. Presumably, human auditory perception is similarly affected by prenatal conditioning.

According to Gibson [1966, p. 5] "the infant does not have sensations at birth but starts at once to pick up information from the world". In other words, infants (like animals) do not have to be conscious of their actions and perceptions in order to learn to interact with their environment, and to become sensitive to perceptual invariances. The same is true for the foetus.

Of all the sounds heard by humans before birth, probably the loudest and/or most consistent is produced by pulsations in uterine blood vessels associated with the mother's heartbeat [Bench 1968; Grimwade et al. 1970], although this sound often falls below the threshold of audibility, which is quite high at low frequencies [Vince et al. 1982b]. The maternal heartbeat may be regarded as an imprinting stimulus on the foetus [Salk 1962]. Prenatal conditioning by heartbeat sounds explains why babies are calmer when fed from the left breast [Lockard et al. 1979]; it may also underlie the emotional connotation of musical *rubato* (tempo fluctuations) [Parncutt 1987b].

Prenatal sounds associated with walking are presumably at least as loud as those of general body movements [Vince et al. 1982b]. Such sounds may underlie strict rhythms, and their association with dance [Parncutt 1987b]. Music therapists [Thaut 1985 and references therein] have exploited auditory rhythms to help disabled children develop fundamental motor skills, supporting the hypothesis that auditory rhythms and body movements are linked through prenatal conditioning. A link between musical tempo fluctuations and prenatally experienced changes in the walking speed of the mother is suggested by experimentally measured timings of musical *ritardandi* [Kronman and Sundberg 1987]. Lerdahl and Jackendoff's [1983] *Metrical Preference Rule 10* – "prefer metrical structures in which at each level every other beat is strong" – is consistent with the theory that rhythm is conditioned by prenatal conditioning by footstep sounds: in asymmetrical positions, the foetus presumably hears one of the mother's feet louder than the other.

The prenatal conditioning hypothesis suggests that rhythmic sensitivity should be associated only with rhythmic *sounds*, and gross bodily movements; pulse-like patterns of visual and tactile sensations should not evoke rhythmic responses. The experimental results of Grant and LeCroy [1986] are consistent with this expectation. They found that performance at rhythmic perception tasks by intellectually disabled people was significantly worse when rhythms were presented in the form of silent taps on the shoulder (i.e. purely tactile stimuli) than when they were presented as drum strokes (purely auditory stimuli), and that the addition of a visual stimulus (allowing listeners to see a drum, or their knee being tapped) did not improve performance.

There is a tendency in music for lower-pitched sounds to be played on the beat, and higher-pitched sounds off the beat. Familiar examples from Western music are the bass/chord alternations of ragtime and waltz accompaniments. In Ghanaian drum music, low-pitched bells are often used at the start of a cycle, and high-pitched bells often sound on offbeats and syncopations. Similarly, in Indian *tabla* music, downbeats are more often played with lower-pitched than higher-pitched drums. Higher pitches *can* also be used to delineate the beat, with low sounds off the beat, but this normally produces a syncopated feel. These observations are consistent with the idea that rhythm is somehow related to heartbeat and walking sounds, as such sounds are concentrated at low frequencies by comparison to other sounds such as the mother's voice. Vince et al. [1982b] reported that intrauterine cardiovascular sounds are concentrated in the range 40–80 Hz.

The foetus is regularly exposed to the mother's speech. A feature of speech to which the foetus might be sensitive is its *intonation*, i.e. rising and falling of pitch. Intonation is developed and mastered by infants much earlier than individual phonemes and words [Tonkova-Yampol'skaya 1973]. The specific and highly structured patterns of Western speech intonation carry much of the meaning of speech [O'Connor and Arnold 1973], in particular, its emotional meaning [Nilsonne and Sundberg 1984].

The musical correlate of speech intonation is melodic *contour*. Apart from rhythm [Jones 1987], contour is the most important aspect of melodic structure for melody recognition [Dowling and Fujitani 1971; Dowling 1978; Massaro et al. 1980]. Contour is more important than conformance to a diatonic scale for the memory of melodies [Davies 1979]: scalar conformance, i.e., the exact sizes of pitch intervals, are "processed separately" from contour [Eiting 1984] and only become important in longer melodies, in which a tonal framework is well established [Edworthy 1985]. Infants, like adults, recognize melodies by their contours, in spite of variations of pitch register and interval size [Trehub 1987], and children learn melodic contour before they learn chromatic interval categories [Dowling 1982; Davidson 1985].

In general, emotion in speech is transmitted by "phonetic content, gross changes in fundamental frequency, the fine structure of the fundamental frequency, and the speech envelope amplitude, in that order" [Lieberman and Michaels 1962, p. 248]. With the possible exception of phonetic content, each of these has a musical correlate. The musical correlate of "gross changes in fundamental frequency" is contour. The musical correlate of the "fine structure of the fundamental frequency" is the amplitude envelope of single musical tones; the fine structure of a single tone can carry a range of emotional meanings [Clynes and Walker 1982]. Musical correlates of speech envelope amplitude are dynamic and accent (i.e., loudness); these, too, are important carriers of emotional meaning in music.

Speech is regularly interrupted by *breathing*. Similarly, musical passages (especially melodies) are organized into *phrases* [Révész 1953], whose durations are comparable with those of breathing (roughly, from 1 to 10 s, including both inhaling and exhaling). This is true not only for music in which phrases correspond to the breathing of the performer (singing, wind instruments) but also in music on instruments which are independent of the performer's breath (strings, keyboards, etc.).

According to Terhardt [1974a, 1976], a prerequisite for the perception of harmony in music is the ability to perceive complex tone sources as single entities, even though the spectrum of a complex tone contains several separately audible components. This ability is supposed to be acquired early in life, as the pitch patterns of the audible harmonics of complex tones become familiar to the auditory system. Experimental evidence on the perception of "missing fundamentals" by infants [Clarkson and Clifton 1985] suggests that this ability is present at the age of seven months. It is probably present at two months: infants of this age perceive speech intonation [Tonkova-Yampol'skaya 1973]. Further experimental evidence suggests that it is present as early as one month after birth: Eimas et al. [1971] found that one-month-old infants categorize speech sounds in much the same way as adults. Eimas et al. concluded that "the means by which the categorical perception of speech . . . is accomplished may well be part of the biological makeup of the organism" [1971, p. 306]. The prenatal conditioning hypothesis provides an alternative (or additional) explanation: these infants had been picking up information from speech sounds not only for the four weeks following birth but also for twenty weeks before birth.

If the ability to perceive complex tones as single entities is acquired prenatally, then there must be a period between the physiological development of the foetal auditory system and birth when the mother's voice is "heard" not as a single sensation but as a simultaneity of five to ten separate sensations corresponding to the lower harmonics (Sect. 6.1.1). This idea adds new meaning to Schenker's [1906] description of the harmonic series as the "chord of nature". Musical chords (or tone simultaneities generally) might be special kinds of "auditory images" [McAdams 1984] associated not with specific environmental sources of sound but with voices heard before birth, especially the mother's voice – the original *corps sonore* (cf. [Rameau 1721], Sect. 1.4.1). This may explain why chord progressions in Western music tend to be preferred to unaccompanied melodies, even though single complex tones are more consonant than chords.

Infants are more sensitive to semitone changes in melodies based on the major triad than to semitone changes in other melodies [Cohen et al. 1987]. According to Trehub [1987, p. 638], "This implies that the diatonic context has greater coherence or stability for infants . . . . It is nevertheless unclear whether diatonic structure in general or major triad tones in particular, with their simple harmonic relations (4:5:6), are responsible for the observed effects. Moreover, one cannot entirely rule out possible contributions of prior exposure to music, however limited." According to the prenatal conditioning hypothesis, the results of Cohen et al. are explicable in terms of prenatal familiarity with the pitch pattern of the 4th, 5th and 6th harmonics of complex tones.

Sounds heard regularly before birth have a calming effect when heard again after birth. This is true for sheep [Vince et al. 1982a] and for humans [Salk 1962]. As we have seen, music includes sounds similar in many ways to sounds heard before birth. This may explain music's wide range of therapeutic applications: it may be used to improve mood and attitude, reduce tension, reduce heart rate and blood pressure, and alleviate psychosomatic illness [Hanser 1985 and references therein]. The prenatal conditioning hypothesis may also explain why music helps learning, especially in learning-disabled students [Gfeller 1983]: subliminal prenatal associations may alleviate anxiety, and promote confidence.

"Music is inherently referential . . . in a very special way it 'means itself'" [Thomson 1983, pp. 4, 5]. With the exception of programme music (e.g. Beethoven's *Pastoral Symphony*), music has little explicit meaning in relation to the human environment, at least not as much as the other arts; its meaning is syntactic rather than semantic [Booth 1981]. Despite this lack of specific associations, music is vivid and powerful. Musical rhythm, melody and harmony express deep emotions and abstract ideas in specific ways [L. B. Meyer 1956; Cooke 1959]. For an individual, this may be due to cultural conditioning, for prenatal conditioning, or both. The nature of cultural conditioning may in turn depend on the role of prenatal conditioning in the historical development of musical styles.

Roederer [1984, abstract] considered that "A most basic issue in the study of music perception is the question of why humans are motivated to pay attention to, or create, musical messages, and why they respond emotionally to them, when such messages seem to convey no real-time relevant biological information as do speech, animal utterances and environmental sounds." He provided three answers to this question, concerning speech acquisition, speech communication and social integration. In a similar vein, Sloboda [1985, p. 266] observed that "it is not at once clear how musical behaviour makes for better adapted individuals that are more likely to survive". He concluded that it is society as a whole (rather than the individual) that benefits from music: "Music, perhaps, provides a unique mnemonic framework within which humans can express, by the temporal organization of sound and gesture, the structures of their knowledge and of social relations" (p. 267).

The Darwinistic approaches of Roederer and Sloboda are similar to the prenatal conditioning hypothesis in that they help to explain why we have music at all. Prenatal conditioning additionally explains the central importance of certain kinds of structure – rhythm, melody and harmony – in music. These structures are cross-cultural, but they are exploited in different ways in different cultures.

How could the prenatal conditioning hypothesis be tested? If prenatal conditioning by heartbeat, footstep and voice sounds has a *direct* influence on musical ability, then the children of bedridden mothers should be less sensitive to rhythm (especially, the relationship between strict rhythms and dance), and children who are born deaf but regain their hearing as a result of an operation (and consequently acquire average hearing ability) should be altogether insensitive to rhythm, melody and harmony. However, such people could also become sensitive to musical structures from *postnatal* experience. In general, the influence of cultural conditioning on an individual's perception of music may be greater than the influence of prenatal conditioning; the effects of prenatal conditioning on music may only reveal themselves in the long term, as musical style develops over generations and centuries. If this is the case, the theory of prenatal conditioning (like Terhardt's theory of pitch, Gibson's theory of direct perception, Darwin's theory of evolution – indeed, *all* theories, according to Popper [1972]) may well be unprovable.

3.2 Consonance

3.2.1 Introduction

Musical sounds are said to be consonant if they are perceived to "sound well" with each other (*con sonare*). An understanding of consonance is essential for an understanding of tonality: the contrast between consonance and dissonance contributes to "tension" and "resolution" [Nielsen 1983] and thereby to a sense of "forward motion" [Forte 1962, p. 15] in tonal music.

The same musical intervals have been described as consonant or dissonant at different stages of Western musical history. For example, the perfect fourth (P 4) was regarded as a consonance in its harmonic function as inversion of the P 5, but as a dissonance in its melodic function as a suspension to the third. A simple modern definition of the (harmonic) consonance of musical intervals states that intervals contained in the major and minor triads (m 3, M 3, P 4, P 5, m 6, M 6, P 8) are consonant, while others (m 2, M 2, TT, m 7, M 7) are dissonant; chords are consonant if they contain no dissonant intervals [Apel 1970]. These definitions apply primarily to simultaneities. In the case of sequential tones, the additional (melodic) role of pitch distance needs to be taken into account (steps are more "consonant" than leaps).

Musical consonance is enhanced by *familiarity* [Cazden 1972]. The more often a relatively dissonant tonal sound or sequence is heard, the more consonant it is judged to be [Heyduk 1975], and discords are objected to less, the more familiar they are [Valentine 1914]. The ability to appreciate the consonance of musical chords is learned by exposure to ordinary complex tones, musical chord progressions, etc. [Terhardt 1974a, 1976]. It seems possible that the ability to appreciate consonance in scales with radically "stretched" intonation could similarly be learned [Pierce 1966; E. A. Cohen 1984].

Consonance also depends on *context*. Chords sound more consonant in familiar contexts (e.g. contexts in which they are diatonic [Gardner and Pickford 1943, 1944; Cazden 1972]). Familiar contexts contain redundancies and therefore carry less information than unfamiliar contexts [Hiller and Isaacson 1959; A. J. Cohen 1962]. Roberts [1986] reported that "Chords presented in a 'traditional' context were judged as being more consonant than the same chords presented in a 'untraditional' context" (p. 169) and "Chords were judged [by untrained listeners] as being more consonant when they are resolved by the following chord than when they are left unresolved at the end of a 'traditional' context" (p. 170). This suggests an expansion of the usual music-theoretical concept of consonance to include sequential (as well as simultaneous) pitch relationships.

In a general approach, sounds at all levels in hierarchies of musical perception may be said to exhibit consonance and dissonance. At the most analytic level, the simultaneous pure tone sensations evoked by a complex tone produced by a musical instrument (string, pipe, etc.) are more consonant than those evoked by a nonharmonic sound such as that of a bell. The simultaneous complex tone sensations evoked by a musical chord are more consonant if the chord has a strong, unambiguous root. The sequential tones of an unaccompanied melody or the sequential chords of a harmonic progression are more consonant if they cover small pitch distances and remain within a standard pentatonic or heptatonic scale. As a rule, consonant sounds are relatively easy to group perceptually; dissonant sounds are not so easily grouped (Sect. 2.3.3).

Consonance should not be confused with *preference*, although in certain cases these terms turn out to be equivalent [e.g. Roberts 1986]. Relatively consonant sounds are not necessarily preferred to relatively dissonant sounds. If this were the case, single tones would always be preferred to chords. Instead, the relationship between dissonance (or complexity, or information flow) and preference (pleasantness, likeability, affective judgment) normally takes the form of an inverted U curve [Smith and Cuddy 1986 and references therein]. For relatively low degrees of dissonance, preference increases with increasing dissonance, while for relatively high degrees, preference decreases with increasing dissonance.

The peak of the inverted U curve is the *optimal dissonance* (or complexity, or information flow) of a piece of music. Degrees of dissonance or complexity which are either lower or higher than the optimum are liked or enjoyed less than degrees near the optimum: music which is too consonant or simple may be boring or irritating, and music which is too dissonant or complex is hard to listen to. Optimal musical complexity depends on the listener, and musical experience [Vitz 1964]. Typically, exposure increases tolerance to (or liking of) dissonance and complexity, causing the peak of the inverted U to move to the right.

Optimal dissonance gradually increased during the history of Western music, as more dissonant sound structures developed [Lundin 1947]. In two-part writing, for example, the generally preferred amount of dissonance changed from that of the fifth in the Middle Ages to that of the third in the Renaissance and later periods. (A notable exception to this was the use of thirds in Medieval England.) The preferred dissonance of chords, with occasional exceptions, increased from that of the major and minor triads in the Renaissance, Baroque and Classical periods to that of more complex chords and tone simultaneities in various romantic, impressionist, jazz, non-diatonic and atonal styles.

In recent decades, this process seems to have slowed or even stopped. Modern audiences prefer different kinds of dissonance in different amounts. Some prefer music with low dissonance (e.g. Baroque and Classical styles), some prefer high dissonance (e.g. atonal styles) and some prefer intermediate degrees of dissonance (late romantics, impressionists, jazz).

Consonance may be divided into *sensory* aspects [Terhardt 1974a, 1976], otherwise known as tonal consonance [Plomp and Levelt 1965], and *cultural* aspects [Lundin 1947; Cazden 1972].

Contributors to the sensory consonance of tone simultaneities include *roughness* and *tonalness* (Sect. 3.2.2). The main culturally conditioned aspect of the consonance of tone simultaneities is the root of a chord (Sect. 3.4.2). Contributors to the sensory consonance of sequential sounds are *pitch commonality* and *pitch proximity* (Sect. 3.2.3). These may be regarded as sensory bases of the culturally conditioned aspects known as harmonic relationship and voice leading (melodic relationship). A further aspect of cultural consonance, involving all lower levels in some way, is the *tonality* of melodies and harmonic progressions (Sects. 3.4.3, 6.2.3–5).

Culturally conditioned aspects of consonance are determined by historical developments. Over the centuries, these, in turn, may be influenced by sensory properties of musical sounds. Many cultural aspects of musical consonance may thus be described as *indirectly sensory*. Indirectly sensory aspects of consonance are conditioned by exposure to aspects of a musical style which appear to have developed as a direct consequence of specific sensory constraints. Although they are sensory in origin, they are only perceived by culturally conditioned listeners.

3.2.2 Roughness and Tonalness

The apparent *roughness* of a sound depends in a complicated way on all its physical properties [Terhardt 1968b, 1974b; Aures 1984]. The simplest case of a rough sound is an amplitude-modulated (beating) pure tone, produced by superposition of two equal-amplitude pure tones of slightly different frequency. The pitch of the tone depends mainly on its *carrier* frequency (the mean of the two original frequencies); the roughness of the tone, on its *modulation* (or beat) frequency (the difference between the two original frequencies). Roughness is evoked by tones with modulation frequencies in the range 20–300 Hz, reaching a maximum at around 70 Hz. Roughness also depends on carrier frequency, but to a lesser extent: it is most pronounced for carrier frequencies near 1 kHz. Contributions to the roughness of a complex sound may be summed over groups of carrier frequencies falling in different critical bands.

An example of a very rough musical sound is a dyad of the complex tones C_4 (middle C) and $D\flat_4$. This dyad contains several amplitude-modulated pure tone components, each formed by the superposition of pure tone components a semitone apart. The fourth harmonics of the two tones combine to produce a waveform with a modulation frequency of 60 Hz and a carrier frequency of 1080 Hz. This waveform, according to the data presented in the previous paragraph, sounds very rough.

In relatively small quantities, roughness can add richness to a sound. For example, full complex tones below about middle C are slightly rough [Terhardt 1974b]: their harmonics are less than 300 Hz apart, so superpositions of neighbouring harmonics have modulation frequencies of less than 300 Hz. This contributes to the rich tone quality of the 'cello. The clarinet, by contrast, has a dark, strangely hollow quality below middle C (in its *chalumeau* register [Kennan 1970]). This is partly due to *lack* of roughness: the even-numbered harmonics of clarinet tones are generally inaudible, so the spacing between odd-numbered harmonics is effectively twice that of the 'cello.

Aures [1984] defined *tonalness* (*Klanghaftigkeit*) to increase as the number and audibilities of pure tone components in a sound increase. In the present study, the term *pure tonalness* is used for this concept. Pure tonalness decreases as the amount of masking among pure tone components increases. It is higher in consonant than in dissonant musical chords, as harmonics of different tones coincide more often, and therefore mask each other less, in consonant chords.

The consonance of musical chords depends partly on the extent to which they evoke complex tone sensations such as "virtual bass notes" [Terhardt 1977]. This is quantified by the parameter *complex tonalness*, the degree to which a sound contains clearly audible complex tone components, i.e. the degree to which it contains clearly audible pure tone components in harmonic pitch patterns. Quantitatively, complex tonalness may be defined most simply to correspond to the audibility of the most audible complex tone component in a sound (Sect. 4.4.3): it is a measure of the ease with which the most salient complex tone sensation in the chord is perceived in the holistic mode of perception, i.e. when no other tone sensation is simultaneously noticed. Complex tones have fewer pure tone components than chords, but they have greater complex tonalness: the pure tone components of complex tones are masked less than those of chords, and are therefore more audible.

Complex tonalness may be enhanced in a musical chord by arranging the tones of the chord so that their fundamental frequencies describe part of a harmonic series. Examples are the major triad and the major-minor (dominant) seventh chord. In this way, complex tonalness influences the chord vocabulary of tonal music (Sect. 6.1.4).

The roughness and tonalness of musical chords are negatively correlated. Roughness is associated with the presence of more than one pure tone component in a single critical band [Plomp and Levelt 1965]. Such components mask each other [Fletcher 1940], reducing tonalness. Sounds in which pure tone components are widely spaced (e.g. octave-spaced tones) have low roughness and high tonalness.

The sensory consonance of a musical chord may be enhanced by playing one note (e.g. the melody) louder than the rest. This reduces roughness and enhances tonalness, as follows. The roughness of a musical chord depends mainly on contributions from pairs of pure tone components of roughly equal amplitude belonging to different complex tones (notes). The roughness contribution from a pair of components falls rapidly as their amplitudes become different, i.e. as the degree of amplitude modulation of the resultant beating tone becomes smaller [Terhardt 1974b]. In a musical chord, roughness may therefore be reduced by playing one note louder than the rest. The same technique enhances the complex tonalness of the chord, simply by increasing the audibility of the most audible tone in the chord. This may explain why it is necessary to play a melody considerably louder than its accompaniment (e.g. on the piano) in order to produce a "singing tone".

3.2.3 Pitch Commonality and Pitch Distance

The present study proposes a new terminology, and a newly structured theory, for the consonance of sequential sound pairs. Like Terhardt's theory of the consonance of tone simultaneities, the present theory has two aspects: pitch commonality, and pitch distance (or its opposite, pitch proximity).

The *pitch commonality* of a sequential pair of sounds is the degree to which the sounds have pitches in common. More precisely, it is the degree to which the sounds evoke tone sensations whose pitches coincide across the two sounds. In categorical perception, pitches up to half a semitone apart or even more may be perceived to coincide in this way (Sects. 2.5.3, 5.6.3).

The pitch of a complex tone is ambiguous (Sect. 2.4.4). Alternative possible pitches lie at important musical intervals (octave, fifth, etc.) above and below the pitch most often heard (the pitch corresponding to the fundamental frequency). So complex tones separated by important musical intervals have pitches in common; they are to some extent "confusable", and therefore possess a certain affinity (*Klangverwandtschaft*) for each other [Terhardt 1983]. This affinity may be regarded as the sensory basis for their consonance.

The idea of pitch commonality is by no means new. In the 19th century, Sechter observed that harmonic relationships between musical chords depended on the presence of common tones, either actual or implied, acting as "harmonic connectives" (*harmonische Bildungsmittel*) or "intermediate fundamentals" (*Zwischenfundamente*) [Watson 1982]. According to this theory, triads of C major and D minor have the tone A in common: the tone A is actually present in the D minor triad, and implied (as a Rameauian *basse fondamentale*) by the C major triad. Helmholtz [1863] thought that musical sounds were harmonically related if they had *frequencies* in common [Cohen 1982]. Western music theory texts recommend that chords in a progression should have *notes* in common, especially in the case of chromatic chords (Sect. 1.1.3). Chords with notes or frequencies in common generally have pitches in common.

The *cycle of fifths* in music theory (Fig. 1.1) may be regarded as a simplified or idealized version of the more general concept of pitch commonality. The pitch commonality of two chords depends on the degree to which they have notes in common, the harmonic relationship between their roots, and their conformance to a particular diatonic scale (Sect. 5.7.3). The pitch commonality of two keys depends on the number of notes the two scales have in common (i.e. the difference between their key signatures, taking into account common additional accidentals such as raised leading notes). Important pitches in musical chords generally coincide with actual notes, or lie at strong harmonic intervals such as octaves and fifths away from actual notes (see Fig. 6.3, Sect. 6.1.5); and diatonic scales are made up of chains of fifths. Consequently, distance around the cycle of fifths reflects harmonic relationships between both musical chords and musical keys.

Small intervals (such as seconds and thirds) are more common than larger intervals in melodies [Jeffries 1974], and chords in a contrapuntal texture "progress" or "lead" better if the voices of the chords traverse small intervals. In a general approach, these effects may be understood in terms of *pitch distance.* In general, the consonance of sequential tones is higher if they are closer to each other on a continuous pitch scale.

The importance of pitch distance is reflected by the importance of the interval of a semitone in the resolution of dissonant or chromatic chords. In the words of Forte [1962, p. 11]:

"The leading note is only one instance of an important melodic law, the *law of the half step.* According to this law the strongest, most binding progression is the half-step progression."

In a similarly vein, Goldman [1965, p. 84] emphasized

"the fact that the semitone gravitates more strongly to its upper or lower neighbor than does the whole tone. The force of the dominant seventh chord is largely the result of the leading-tone's drive to the tonic and the fourth's drive to the third."

Pitch distance is difficult to define quantitatively in the case of pairs of complex sounds such as musical chords. In analytical perception, several tone sensations (mostly corresponding to actual notes) are noticed at the same time in each chord. The overall apparent pitch distance between two chords may be assumed to depend in some way on the pitch distances between all sequential pairs of tone sensations (Sect. 4.6.2).

3.3 Musical Pitch

3.3.1 Octave Equivalence

Octave equivalence is a musical universal [Hardwood 1976]. Musics from all over the world, including many non-Western musics (see e.g. Nettl 1956], exhibit octave equivalence: notes an octave apart are interpreted by musicians as similar or identical in harmonic function, and are given similar or identical names. Octave equivalence is "one of the most fundamental axioms of tonal music" [Forte 1962, p. 7], allowing the same scale degree (e.g. the tonic) to be played in all pitch registers, and by all pitched instruments and voice types.

Octave equivalence aids in composition, performance and improvisation. It reduces the number of different compositional possibilities, and so makes music easier to organize and remember. Pitch terminology is simpler when notes an octave apart are called by the same name. Notation would also be simplified if notes an octave apart looked similar on paper: surprisingly, this possibility is not used to its full advantage in Western notation [cf. Read 1987].

The sensory basis for octave equivalence has been described by Terhardt [1972, 1974a] and further examined in the experiments of the present study (Chap. 5). *Simultaneous* complex tones an octave apart blend so well that the result sounds almost like a single tone, because all harmonics of the upper tone coincide with even-numbered harmonics of the lower tone. Single complex tones contain prominent pitches an octave apart, so *sequential* tones an octave apart have many pitches in common and are therefore more or less "confusable" with each other [Terhardt 1983] (Sects. 5.4–6).

Of the above, the simultaneous effect implies a greater degree of "equivalence" [Deutsch 1982a]. Transposition of one of the voices of a chord through an octave can make very little difference to the sound of the chord. By contrast, transposing a note in a melody through an octave changes the character of the melody altogether, especially if the melody's contour is affected. So melodies in which both pitch class and contour are maintained but pitch register is not (i.e. tones appear in the wrong octave) are difficult to recognize and recall [Deutsch 1972b; Dowling and Hollombe 1977; Deutsch and Boulanger 1984].

In a psychoacoustical approach, the difference in consonance between the octave and the fifth is only one of degree [Terhardt 1976]. The singular and categorical importance of the octave in music is not immediately obvious from the pitch properties of complex tones and intervals. Perhaps the main reason why the octave, and not some other interval, became *the* equivalence interval in music is that *combinations* of octaves are consonant. The fifth could not be used as an "equivalence" interval in the way that the octave is, as combinations of fifths (ninths, thirteenths, etc.) are much less consonant than combinations of octaves.

3.3.2 The Chromatic Scale

A musical scale consists of discrete pitches or pitch categories separated by specific intervals. It is a framework within which tonal music may be composed, performed and improvised [Pressing 1978; Terhardt 1979b]. The historical development of musical scales is influenced, but not predetermined, by the physical properties of tones [Cazden 1954; Shirlaw 1957].

The chromatic scale has provided a basis for the composition of tonal music in the West since all twelve scale degrees came into common usage in the late Middle Ages. The heptatonic modes and diatonic scales may all be regarded as subsets of the chromatic scale.

The chromatic scale need not be equally tempered [Rasch 1983]: any standard heptatonic scale with provision for five extra notes per octave, so that the resultant twelve notes are roughly equally spaced and every note has a reasonably tuned upper and lower perfect fifth, may be regarded as chromatic. In modern practice, the most common or average tuning of the chromatic scale is that of *equal temperament*, in which each semitone covers the same distance on a frequency level (or log frequency) scale. The principle of twelve-tone equal temperament was first clearly articulated in the Europe by Mersenne in 1635 [Apel 1970]; its invention over a thousand years earlier by the Chinese had been hidden from the West by the language barrier [McClain 1979]. Bach promoted equal temperament in keyboard instruments with the publication of the two parts of his *Wohltemperiertes Klavier* in 1722 and 1744.

The chromatic scale is cross-cultural, but not universal. It was in use in ancient China [McClain 1979] and later evolved independently in India and Persia [Burns and Ward 1982], before emerging in the West. More recently, it has spread from the West to other cultures [Nettl 1986], although largely for political and social (rather than intrinsically musical) reasons. These observations suggest that the chromatic scale has a sensory basis – that its development both in the West and elsewhere was influenced by universal sensory properties of complex tones. This conclusion is not supported by simplistic treatments based on the harmonic series (Sect. 1.2.3): more sophisticated musical and psychoacoustical explanations are required.

The chromatic scale may be generated by successively traversing rising fifth intervals in one direction and octaves in the other (the cycle of fifths). After a total of twelve fifths and seven octaves, a coincidence occurs. If these intervals are exactly tuned to frequency ratios of 3 : 2 and 2 : 1 respectively, the coincidence is not exact: the start and end points are 0.24 semitone (one "Pythagorean comma") apart. However, this distance is negligible by comparison to the cumulative uncertainties in the sizes of the twelve fifths and seven octaves, given that the sizes of fifths and octaves are generally uncertain by 0.05 – 0.1 semitone, even in excellent musical performances [Sundberg 1982]. In musical practice, then, the cycle of fifths is closed and complete.

The major-third interval has played an important harmonic role in Western music since the sixteenth century (earlier in England). The lower note of the major third, like that of the perfect fifth, often corresponds to the root of a chord. According to Terhardt, this is due to familiarity with the pitch pattern of complex tones: the perfect fifth occurs between the 2nd and 3rd harmonics, the major third occurs between the 4th and 5th harmonics, and the 2nd and 4th harmonics are octave-equivalent to the fundamental. Taking the harmonic role of the major third into account, the chromatic scale in diatonic music may also be generated by three intervals: the octave, the perfect fifth and the major

Fig. 3.1. Generation of the chromatic scale by combining octaves, fifths and major thirds. *Open noteheads*: standard heptatonic scale (generated by fifths and octaves). *Full noteheads:* non-heptatonic (chromatic) notes

third [Schoenberg 1911; Shirlaw 1957; Longuet-Higgins 1979]. In this approach, the standard heptatonic scale (the basis of the medieval modes, later specified by conventional key signatures) may be generated by octaves and fifths, while other, non-heptatonic notes lie at major-third intervals away from heptatonic notes (Fig. 3.1).

This derivation of the chromatic scale explains only the *harmonic* function of each non-heptatonic note, in a way which corresponds to its enharmonic spelling. *Melodic* functions and enharmonic spelling depend primarily on semitone relationships, i.e. pitch distance (e.g. C# → D, D♭ → C). The enharmonic spellings G♭ and A# are absent from the figure, as they have no harmonic function in C major or A minor. These notes may nevertheless have melodic functions, as chromatic auxiliaries to F and B respectively.

Twelve-note equal temperament approximates the small-number frequency ratios of musical intervals (especially the fifth) remarkably closely. Other equal temperaments with this property include (amongst others) 19-, 31- and 53-fold subdivisions of the octave [Yasser 1929; Fokker 1966; Pikler 1966; Hoerner 1976; De Klerk 1979; Yunik and Swift 1980; see also Balzano 1982). None of these has succeeded in materializing into new harmonic styles. The following psychoacoustical analysis suggests that this failure is due to more than just musical conservatism.

Harmonics of a complex tone which are mistuned by up to half a semitone are still perceived to belong to the tone; harmonics mistuned by a whole semitone or more are not [Moore et al. 1985]. This may be understood in terms of familiarity with the spectral pitch patterns of complex tones. The exact sizes of the intervals between spectral pitches (octave, fifth, fourth, . . .) vary due to pitch shifts [Terhardt 1979a] and, sometimes, due to physical deviations from harmonicity (e.g. in pianos [Schuck and Young 1943]). These variations must be accommodated by the auditory system if complex tone sources are to be consistently recognized. Given that the root of a chord is a virtual pitch associated with a harmonic pattern of spectral pitches from different musical notes [Terhardt 1974a] (Sect. 3.4.2), division of the musical octave into 19 or more parts would allow adjacent scale steps to contribute to the same root. Thus, adjacent scale steps would no longer have the distinct harmonic functions that they have in the twelve-tone system.

The above argument ignores the role of roughness, and hence of exact frequency ratios, in harmony (Sects. 1.2.4, 3.2.2). In Indian *sitar* music, roughness appears to have a strong influence on intonation, as melody tones and their harmonics beat with the harmonics of the sustained drone. This may be why interval gradations (intonations?) of less than a semitone have structural significance in Indian music theory (although this is not necessarily the case in practice, see Jhairazboy and Stone [1963]).

Musical scales normally have less than twelve notes per octave, perhaps due to limitations of perceptual information capacity [Miller 1956]. Commonly used subsets of the chromatic scale are the standard pentatonic scale (e.g. in China) and standard heptatonic scale (e.g. in the West and in India). They are common because of their harmonic coherence (they describe unbroken sections of the cycle of fifths); because they each contain only two interval sizes and no adjacent half steps [Pressing 1978]; and because of their asymmetry, which prevents them from being mapped onto themselves by transposition within the octave [Balzano 1980; Cross et al. 1985].

World musics use a wide variety of other musical scales which may be regarded as subsets of the chromatic scale (when defined as a set of pitch *categories*, not exact pitches). In addition, there are scales which do not easily fit into the chromatic mould, e.g. in Indian, Thai and Indonesian music. Such scales may be based on quite different sensory and cultural premises from those which underlie the standard pentatonic and heptatonic scales.

3.3.3 Intonation

The three major theoretical intonations (tunings) in Western music are *Pythagorean* (based on frequency ratios containing only powers of two and three), *just* or *pure* (in which powers of five also appear) and *equal temperament* (in which semitones correspond to equal frequency ratios). In Pythagorean temperament, major intervals are wider, and minor intervals narrower, than in equal temperament; the reverse is the case for just intonation. For example, an equally tempered major third spans exactly 4 semitones, a Pythagorean major third (with a frequency ratio of 81 : 64) spans 4.08 semitones, and a just major third (5 : 4) spans 3.86 semitones.

The differences between these tunings are generally imperceptible to non-musicians [Roberts 1986] and are smaller than typical variations in intonation even in excellent performances [Seashore 1938; Burns and Ward 1982]. There is nevertheless an extensive literature on the subject of intonation, perhaps because many researchers believe (at least at the outset of their research) that the musical meaning of an interval is somehow embodied by its frequency ratio (Sect. 1.2.2), and that the "actual" frequency ratios of musical intervals revealed by intonation research might lead to a better understanding of harmony and tonality.

The optimal intonation for sustained harmony (simultaneous tones) is often found in experimental studies to lie on the "just" side of equal temperament [Norden 1936; O'Keefe 1975; Roberts and Matthews 1985], although sometimes equal temperament is preferred to just [e.g. Roberts 1986]. Just temperament causes harmonics to coincide, minimizing beats [Helmholtz 1863] and thereby roughness [Terhardt 1968 b].

The musical instrument which is perhaps most susceptible to the dissonant effect of beats produced by tempered tuning is the pipe organ. The introduction of equal temperament in pipe organs was resisted until long after it had become the norm in other keyboard instruments such as the piano; in England, for example, it was not widely accepted until the nineteenth century [Williams 1968].

Just intonation only applies when individual complex tones are exactly harmonic. Piano tones, for example, are not exactly harmonic: the frequencies of their pure tone components are slightly stretched relative to a harmonic series [Ward and Martin 1961]. Further, the "exact frequency ratio" principle only applies when there is no vibrato. Even in barbershop singing, where vibrato is not normally noticeable, variations in intonation of 0.1 – 0.2 semitones may fail to give rise to beats and roughness: the periods of time over which individual voices are periodic, and hence exactly harmonic, are limited [Hagerman and Sundberg 1980].

Intonation of *melodies* in Western music tends to the Pythagorean side of equal temperament, especially in violin playing [Greene 1937; Nickerson 1949; Cazden 1954; Fransson et al. 1974). Pythagorean intonation emphasizes the difference between major and minor intervals, and tends to "anticipate" resolution of dissonance through semitone steps. Since beats play no role in the perception of sequential tones (Sect. 1.2.4), an explanation of quasi-Pythagorean intonation of melodies in terms of frequency ratios (such as 81 : 64 for the major third) is far-fetched [cf. Ward 1962].

The distinction between melodic and harmonic intonation is by no means clear cut. In one study [Shackford 1961, 1962] measurements were made of frequency ratios between tones in string trio performances. No systematic difference was found between the intonation of simultaneous and sequential tone pairs. Similarly, in a test involving intonation of both melodic and harmonic intervals by 48 wind players, Duke [1985] found that, overall, there was no consistent tendency to play either sharper or flatter than equal temperament.

The intonation of some scales and melodies, e.g. in Swedish folk music [Tjernlund et al. 1972], shows the opposite tendency to that normally found in the melodies of classical Western music: major intervals are played smaller than equal and just temperament, and minor bigger. In this way, the spacing of the seven tones of the standard heptatonic scale becomes more uniform relative to their chromatic (equally tempered) spacing. Roughly equally spaced scales (including 12-note equal temperament) have the advantage that expressive and random deviations from the centre of a scale step (pitch category) may more easily be accommodated without risking confusion with neighbouring steps.

Experiments on intonation are complicated by the effect of *pitch shift* [Terhardt 1972, 1974a]. The pitch of a pure tone or a pure tone component depends on its intensity [Stevens 1935] and on the simultaneous presence of other sounds which mask the tone [Egan and Meyer 1950; Webster et al. 1952; Terhardt and Fastl 1971]. Both these effects are related to masking patterns, i.e. to the shape of the masked threshold [Webster and Schubert 1954; Hesse 1987]. Pitch shifts of pure tones can amount to as much as a semitone, or even more than a semitone. The pitch of a complex tone or complex tone component (e.g. in a musical chord) is more stable with respect to changes in level and masking [Lewis and Cowan 1936; Schouten 1940; Stoll 1985].

The intonation of sequential complex tones in music is affected not so much by the pitch shift of the main tone sensation evoked by each tone as by shifts in the sizes of the *intervals* between the main tone sensation and subsidiary tone sensations, especially those at the upper and lower octaves. These intervals are slightly "stretched" due to pitch shifts [Terhardt 1970, 1972, 1974a]. Given that sequential tones an octave apart are tuned by lining up near-coincident pitches (i.e. maximizing pitch commonality), this explains why the octave interval between sequential complex tones in music tends to be tuned to a frequency ratio slightly larger than 2:1 [Corso, 1954; Terhardt 1971; Sundberg and Linqvist 1973; Terhardt and Zick 1975]. Octave stretching is also found between sequential pure tones [Ward 1954; Walliser 1969a; Terhardt 1970].

According to the above authors, perceptually optimal octave intervals between sequential pure and complex tones correspond to frequency level differences of 12.1 – 12.4 semitones, depending on pitch register. A stretch of about 0.15 semitones is typical between sequential tones in the middle pitch register [Fransson et al. 1974]. Intonation of simultaneous complex tones normally involves a compromise between the requirements of *pitch*, by which octaves tend to be stretched, and *roughness*, by which octaves are normally "pure", i.e. not stretched [Terhardt 1974a, 1976]. On average, it may be assumed that musical intervals are stretched by about 0.1 semitone per octave.

Because of octave stretch, octave-periodic scales and intonations (including equal temperament) tend to be slightly stretched throughout. This is especially true in the case of the piano, whose tones have stretched harmonics [Ward and Martin 1961]. Hence the results of experiments in which performed sequential intervals are found to be tuned consistently larger than equal temperament are explicable either in terms of familiarity with piano tuning (a cultural explanation) or in terms of optimization of pitch commonality (a sensory explanation). On instruments without octave stretching (e.g. electronic and pipe organs), very high tones can sound flat by comparison to very low tones, so much so that large intervals can be miscategorized: for a listener used to a stretch of 0.1 semitone per octave, an interval of five octaves could be heard as four octaves and a major seventh. The extent of this effect depends on the individual pitch shifts of the tones, which in turn depend on their level and spectral composition.

Small intervals are sometimes found to be tuned consistently smaller than equal temperament. This effect has been reported in the case of minor seconds [Ward 1970] and seconds and thirds [Rakowski 1976]. It presumably enhaces pitch *proximity*, without affecting chromatic pitch and interval categories.

3.4 Tonality

3.4.1 Introduction

In a general definition, the concept of tonality "embrances the main structural components of the tonal composition; within it are expressed the highly diversiefied events and multiple relationships which form a ... musical unity" [Forte 1962, p. 79]. In this view, tonality includes all aspects of consonance described in the previous sections.

More specifically, tonality may be defined as the perceptual organization of a passage of music around a tonic or key. Music theory is not very specific about what "tonic" and "key" mean, however. Berry [1976, p. 22] regards the tonic as "a specific pitch-class-complex of resolution", i.e. as a pitch class, dyad class, chord class or scale. Similarly, Wilding-White [1961] regards tonality as a function of a pitch set, which could be a tone, a chord or a scale.

The kind of musical element around which tonality revolves depends on the kinds of musical element making up the passage of music in question. The tonic of a *melody* may be regarded as a specific pitch or note [Erickson 1984]. More often, octave equivalence is assumed, and the tonic of a melody is regarded as a pitch class (e.g. "C" instead of "middle C"). The tonic of a *chord progression* may be regarded as a specific (root position) chord, often appearing at the start and/or the end of the progression. Given octave equivalence, however, it is more common to specify the tonic as a *chord class*: as a set of pitch classes (e.g. CEG) rather than a set of actual pitches (e.g. $C_3 G_3 C_4 E_4$). Similarly, the root of a chord is normally specified not as an actual pitch but as a pitch class.

Diatonic tonality arose in Western music soon after composers began to treat the major and minor triads as consonant and harmonically functional (around the sixteenth century). This suggests a link between the tonic of a diatonic scale and the root of a chord.

3.4.2 The Root of a Chord

The root of a chord may be defined most simply as the note or pitch after which the chord is named in Western music terminology and theory. According to this definition, the root of C–E–G is C simply because this chord is called "C major". In music theory, the root may be regarded as that single bass note which most satisfactorily represents the function of a chord in a harmonic progression. If the chord C–E–G in a chord sequence were to be replaced by a single bass note, the note which would least disturb the harmonic progression would be C. Roots are normally understood to be independent of the voicing (including inversion) and relative accentuation of the notes of a chord.

A chord's root determines its diatonic function [cf. Riemann 1893]. The concept of the root is therefore essential for the understanding of Western harmony. Yet despite centuries of music-theoretical development, there still exists no widely accepted theory of the root's nature and origins.

Rameau [1721] compared the notes of a musical chord with the harmonics of a single complex tone (such as a typical musical tone) and hypothesized a correspondence between the root of a chord and the fundamental of a harmonic series of frequencies. During the nineteenth century, Rameau's approach was regarded as the basis for a "vertical" view of harmony, complementing the "horizontal" view based on the diatonic scale and figured bass [Watson 1982]. Theorists such as Bruckner tended to place more emphasis on the "vertical" aspect of harmony, asserting that not only triads and sevenths but also ninths were explicable in terms of Rameau's theory. Others [e.g. Schenker 1906] emphasized the "horizontal" aspect, claiming that only triads were derived from the harmonic series ("Nature") while sevenths and ninths were produced by melodic or horizontal elaboration of triads (by the "Artist").

Music theories based on frequency ratios and the harmonic series have never been able to explain satisfactorily the nature and root of the second most common chord in mainstream Western music, the minor triad (Sect. 1.2.3). The most commonly advocated frequency ratio representation of the minor triad, 10:12:15, suggests that the root of the chord lies a major third below the conventional root, i.e. that the triad is part of a major seventh chord (8:10:12:15). Furthermore, the theory suggests that the minor triad is considerably more dissonant than the major; in musical practice, however, these two triads are about equally consonant.

A commonly proposed solution to this theoretical dilemma was the theory of *harmonic dualism*. Theorists such as Zarlino, Rameau, Tartini, Hauptmann and Riemann advocated that "since major and minor triads were supposed to produce opposite psychological effects on the listener they must therefore be based on opposite principles" [Jorgenson 1963, p. 31]. Accordingly, they proposed that the major triad was based on harmonics 1, 3 and 5 (or 4, 5 and 6), while the minor triad was based on the corresponding *subharmonics* (or *harmonic undertones*). Thus the single note C generated both the C major triad (through its overtones) and the F minor triad (through is undertones).

The theory of undertones has some serious faults. It still fails to predict the root of the minor triad – the root of F minor isn't C! And the underlying theory itself is implausible: subharmonics are not found in naturally occurring sounds.

According to the *phonic theory of reinforced resonance* of Oettingen and Helmholtz [Jorgenson 1963; Christensen 1987], the tones of the minor triad each have one harmonic in common. For example, G_6 is the 6th harmonic of C_4, the 5th of $E\flat_4$, and the 4th of G_4. Again, this does not explain why the root of C minor is C. Nor, for that matter, do combination tones (Sect. 1.2.5).

The music-theoretical root of the minor triad may be correctly predicted by treating it as a distorted form of the major – a major triad with a lowered third, or with a different mediant dividing the fifth. Advocates of this idea included Rameau, Hindemith, Tovey and Schenker [Jorgenson 1963]. The trouble with this approach is (again) the implication that the minor triad is categorically inferior to or less fundamental than the major.

In a psychoacoustical approach, all chords, including the major and minor triads, are initially treated simply as tone simultaneities, i.e. as sounds composed of tones which happen to occur at the same time. The ability to perceive a chord in this way is assumed to be acquired by familiarization with complex tones in the auditory environment.

According to Terhardt [1974a, 1982], the root of a chord is a virtual pitch, i.e. a complex tone sensation. This observation alone is not very useful, as there are virtual pitches at *all* the notes of the chord. The root is different in that the spectral pitches in its harmonic pitch pattern arise from more than one complex tone (note). In other words, the root is the implied fundamental of a group of pure tone components belonging to different complex tones. For example, the root A of the chord A–C–E may be produced by pure tone components at A (from the note A), at E (from the notes A, C or E), at C# (from the note A), at G (from A or C) and at B (from A or E). In a given chord, there are generally several complex tone sensations of this kind: the root is *ambiguous.*

The perception of a chord as a tone simultaneity is appropriately modelled by psychoacoustical methods based on universal aspects of hearing (Chap. 4). This enables specific predictions to be made concerning the roots of musical chords. The music-theoretical root of any chord class in mainstream harmony theory (including the minor triad) turns out to be the pitch class with the highest calculated salience (Sect. 6.1.6); and the root of a chord in a specific voicing generally corresponds to the most prominent pitch class in the bass region (Sect. 6.1.5).

The pitch pattern of a complex tone may be recognized even if parts of the pattern are missing, or extra elements are added (Sect. 2.3.3). Similarly, in a psychoacoustical approach, the root of a chord may still be perceived if notes corresponding to harmonics of the root are missing, or if notes not corresponding to harmonics are added. For example, the root of the C major triad is weakened, but not changed, if the note E (which corresponds to the fifth harmonic of C) is replaced by E♭ (which doesn't correspond to any normally audible harmonic of C); the root (C) is maintained by the strong root implication of the fifth C-G.

A mere tone simultaneity has limited "meaning" in a musical context. To be musically meaningful, a chord must be perceived not only as a tone simultaneity but also as a *musical element.* The ability to perceive a chord in this way is acquired by cultural conditioning (Sect. 3.1.1). It appears to involve an awareness of the harmonic nature of pitch intervals between the noticed tone sensations of the chord, and of the relationship between noticed tone sensations and the root of the chord.

3.4.3 The Tonic of a Scale

Melodies from all over the world exhibit tonality in that they centre on a particular "tonic" [Erickson 1982]. The most universal way of "tonicizing" a pitch is by direct emphasis: repetition, duration and accentuation. Another apparently universal effect is that the tonal centre tends to be near the centre of the pitch range or tessitura [Erickson 1984], so that the average pitch distance between the tonic and other scale notes is as low as possible. Both these sensory affects influence the tonics of world scales, including the "finals" of church modes in Western music of the Middle Ages.

By contrast, the tonic of a *diatonic* scale is analogous to the root of an elaborated broken chord. There is no sensory basis for the root of a broken chord: pure tones must be simultaneous, or very close to simultaneous, to form complex tone sensations [Hall and Peters 1981]. Yet both children and adults with experience of Western harmony extract the tonal hierarchy of a major scale from the sound of a broken major triad [Cuddy and Badertscher 1987]. This suggests that the perception of the root of a broken chord can develop by exposure to unbroken chords with clear roots, such as major triads, minor triads and major-minor ("dominant") sevenths. Such chords were first regularly heard in Western music in the late Middle Ages and Renaissance; diatonic tonality came into being soon after. Another musical development just preceding the emergence of diatonic tonality was the use of perfect or V-I cadences. As these began to influence tonicization, hierarchical nature of diatonic tonality began to depend on root relationships between notes a fifth apart [Schenker 1906].

The emergence of diatonic tonality reduced the number of theoretically available scales from seven (modes starting on different degrees of the standard heptatonic scale) to two (the major and minor scales). In practice, the change was not as great as implied by these figures, as some of the modes of Medieval theory were rarely used, and the melodic minor scale had different ascending and descending forms. But it certainly was significant: it paved the way for the drama of the classical sonata form, and the expressive possibilities of romantic harmony.

The tonic of a diatonic scale turns out to be the *root of the chord formed by scale notes not involved in tritone relationships* [Parncutt 1987a]. Tritone relationships eliminate B and F from C major, and B, F, D and A*b* from C harmonic minor. The remaining notes (C, D, E, G and A in C major, and C, E*b* and G in C harmonic minor) are tonally more stable than the eliminated notes, which in turn are more stable than non-scale (chromatic) notes. The root of the chord C–D–E–G–A is ambiguous: its most important root is C (hence the chord symbol C^{6add9}), its subsidiary root, A (A^{m7add4}). This explains why the tonic of the C major scale is usually C, and occasionally A (in the case of the descending melodic minor scale on A). The tonic of the chord C–E*b*–G is clearly C. The tonic is also the lowest note in a chain of fifths formed by the more stable scale degrees: the tonic of C major is the lowest note in the chain C–G–D–A–E, and the tonic of C minor is the lowest note of C–G (the only fifth among the notes C, E*b* and G).

The above procedure is based on sensory aspects of musical tones; but the application of the procedure to the determination of tonics is clearly specific to the Western diatonicism. In the words (and italics) of Cazden [1954, p. 290], "The natural potentials of tone act as a *limiting condition* of the art of music, but not as a *determining cause* of musical relations." The tritone is the least tonal interval class (not counting the major seventh, which is the inversion of the minor second): it combines low harmonic relationship with low pitch proximity. The avoidance of tritones dates from Medieval times; note, however, that it was still possible in the Middle Ages for a tritone to be included above the final (e.g. in the Lydian mode FGABCDEF). The importance of chord roots for diatonic tonality dates to the Renaissance, when major and minor triads began to be used as musical elements (rather than as more or less coincidental tone simultaneities in Medieval counterpoint); and the importance of the perfect fifth interval dates to the emergence of the V-I cadence at roughly the same time. As explained in the last section, the root of a chord has a sensory basis, and the fifth is the interval which has the strongest influence on the root of a chord.

Musically useful subsets of the chromatic scale seldom have two semitones in a row, or "adjacent half-steps" [Pressing 1978]. The only single chromatic alterations to the scale CDEFGAB which do not produce adjacent half-steps are C#, E♭, F#, G#, A♭ and B♭. According to the above procedure, the tonic of the scale C# DEFGAB is the root of the chord D–E–A. The most likely root of this chord is D, as both E and A correspond to harmonics of D; A is a subsidiary root, as D does not correspond to a harmonic of A. Applying this procedure to all the above chromatic alterations produces the following predicted tonics: are D, C, G, A, C and F (respectively). Thus, only C#, F#, G# and B♭ suggest modulations to new keys, while E♭ and A♭ suggest instead a shift toward the tonic minor.

In this light, the harmonic minor scale may be regarded as a chromatic alteration of the major in which only those notes which are not crucial for the determination of the tonic have been flattened. Conversely, the major scale may be regarded as a chromatic alteration of the harmonic minor, in which only those notes which are not crucial for the determination of the tonic have been sharpened. Major and minor scales on the same tonic may thus be regarded as belonging to one and the same "major/minor" tonality [Goldman 1965].

In the above examples of chromatic alterations to the standard heptatonic scale, the notes C#, F# and G# function as *leading notes*. Leading notes have an important *melodic* function: they lie at semitone intervals away from tonally strong pitches (here, D, G and A respectively). In the light of the proposed procedure for the determination of the tonic of a scale, rising leading notes (such as these three) also have a *harmonic* function: their tritone relationship with the subdominant (scale degree IV, the lowest note in a chain of fifths formed by *all* scale notes) prevents the subdominant from becoming the tonic. Falling leading notes have no such harmonic function; this may explain why they are much less common in diatonic music than rising leading notes.

The fifth C–G in the key of C major/minor is "authentic" in the sense that its root corresponds to the tonic. The fifth F–C is "plagal": its upper note (not its root) is the tonic [Shirlaw 1957]. The status of the F–C fifth may be changed to "authentic" by replacing the scale note B by B♭. This eliminates the tritone relationship against the F (making it a tonic candidate) and creates a new tritone between B♭ and E (preventing B♭, the lowest element in the new chain of fifths, from being the tonic).

Sharps relative to a prevailing key signature lie at major-third intervals above prevailing scale notes; flats, at major-third intervals below scale notes (Fig. 3.1, Sect. 3.3.2). The root-implying property of the major-third interval means that flats relative to a prevailing key signature are more likely to act as roots of chords than sharps. In other words, flats are more *salient* than sharps. This explains the finding of Thompson and Cuddy [1986] that modulation to flat keys is more noticeable than modulation to sharp keys, i.e. adding *n* flats to (or removing *n* sharps from) a key signature produces a more distant modulation than adding *n* sharps (or removing *n* flats).

The diatonic scales may be regarded as the main musically conditioned aspect of the perception of Western melodies (just as streaming is its main sensory basis, Sect. 2.4.6). The diatonic scales are so familiar that it is useful to model melodic perception by matching the pitches of a melody to a rigid *diatonic template* [Jordan and Shepard 1987] (Sect. 2.4.3). This approach is consistent with the following experimental data: listeners quickly recognize fragments of diatonic scales in melodies, and expect subsequent tones to belong to the same scale [Francès 1972; Deutsch and Feroe 1981]; well-structured melodies (i.e. tonal melodies) are remembered more easily than atonal melodies [Dewar et al. 1977]; and nondiatonic tones in diatonic melodies are recognized less often than diatonic tones [Krumhansl 1979].

The rarest interval in the major scale is the tritone: it occurs only between scale degrees VII and IV. This allows the major scale (and therefore its perceptual "template") to be specified fully and unambiguously by two notes a tritone apart and one other note [Browne 1981]. This set of three pitches is therefore sufficient for the recognition of the scale and its tonic [Butler and Brown 1984]. For example, the notes B, F and any other "white note" (C, D, E, G, A) are only diatonic in the key of C major. The notes B, F and a "black note" could only belong to F♯/G♭ major. These considerations do not apply for the harmonic minor scale, as it includes two tritones (II–VI as well as VII–IV).

3.4.4 Major/Minor and Emotion

The major/minor distinction has been central to the emotional meaning of diatonic music at least since the Renaissance [Wienpahl 1959]. Other things being equal (tempo, timbre, texture, etc.), music in major keys is supposedly appropriate for the expression of positive emotions (happiness, brightness, confidence, victory, . . .) and music in minor keys expresses negative emotions (sadness, darkness, defeat, tragedy). There is no doubt that Western listeners are sensitive to the emotional connotations of major and minor: recent research has gone so far as to demonstrate a connection between preference for minor tonality and oral dependency, i.e. a longing to be supported and nurtured [Juni et al. 1987]. However, the *origins* of the emotional connotations of major/minor are by no means clear.

Almost by definition, the "majorness" or "minorness" of a piece of music is determined by the "majorness" of "minorness" of intervals above the tonic. The most important interval in this regard is the third; the second most important, the sixth. It follows from this that the tonal structure of a piece needs to be firmly established before the piece can take on an unambiguously "major" or "minor" flavour. As yet, experimental studies on the emotional meaning of intervals [Maher 1980] and on the perceived major/minorness of triads [Crowder 1985] have not involved clear, unambiguous tonal contexts, and so have thrown little light on the emotional meaning of specific tonal progressions according to music theorists [e.g. Cooke 1959].

Musical styles develop under the continuous influence of sensory (universal) aspects of musical sounds (Sect. 3.1.1). The following sensory aspects are relevant for their emotional meaning: (i) the emotional meaning of pitch patterns in speech, and hence in melodies [Lieberman and Michaels 1962; Nilsonne and Sundberg 1984], and the emotional meaning of subtle pitch changes in individual musical notes [cf. Clynes 1977; Senju and Ohgushi 1987]; (ii) the presence of the pitch pattern of the major triad in the pure tone sensations of a single complex tone; and (iii) the dissonance and root ambiguity (multiplicity) of the minor triad, which is slightly greater than that of the major (Sects. 6.1.3, 6.1.4). One or more of the above three factors could contribute indirectly to the "happy/sad" (or "equilibrium/tension", or "pleasure/unpleasure") aspects of major/minor in Western music.

(i) According to Forte [1962, p. 11], "If both whole-step and half-step progressions are available from a given note the half-step progression will be preferred." This explains the melodic function of the leading notes of diatonic scales. It also explains the tendency of the third step of the major scale to rise to the fourth, and of the third and sixth steps of the minor scale to fall to the second and fifth respectively. This suggests that the emotional meaning of major/minor may be associated with rising and falling of pitch.

A problem with this approach to the understanding of major/minor is that the relationship between the intonation and emotional meaning of speech may be too complex to permit such simple generalizations. Intonation of statements, question, imperatives and exclamations are similar in different languages [Tonkova-Yampol'skaya 1973] but there are also important differences, even within languages (dialects and regional variations).

(ii) The "major" pitch pattern of the pure tone sensations of a complex tone could contribute to the emotional meaning of major/minor via prenatal conditioning (Sect. 3.1.2). If isolated speech vowels heard by the foetus sound like musical chords at some stage of prenatal development, major triads heard after birth may be associated more strongly with the security of prenatal experience than minor triads or other chords.

(iii) The dissonance and root ambiguity of the minor triad may have gradually contributed to the emotional meaning of major/minor during the historical emergence and development of diatonic tonality. Major triads more clearly and positively define the tonic of a passage than minor triads [Cuddy and Lyons 1981]; the "sadness" of minorness may be associated with a feeling of uncertainty or indecisiveness about the position of the tonic in minor keys.

The emotional characteristics of the Medieval church modes and ancient Greek modes were rather different from those of the diatonic scales [Révész 1953]. This is not surprising: these modes were used only for melody and polyphony, in a weak tonal context. They were not used for triadic harmony. The above three hypothetical sensory contributions to the emotional meanings of intervals apply only to chord progressions, or to melodies and contrapuntal passages which imply chord progressions.

3.4.5 Chord Progressions

The consonance of a chord progression depends on: the consonances of individual chords in the progression (Sect. 3.2.2); the consonances of pairs of chords, i.e. how well they are perceived to go with each other (Sect. 3.2.3); the unifying effect of melodic streaming (Sect. 2.4.6); and the strength of the tonal structure (tonality) of the progression (Sect. 3.4).

These contributions to the consonance of a progression are normally positively correlated with each other: progressions with strong tonal structure often contain consonant chords (such as the tonic), consonant chord relationships (such as those between the tonic and other chords), and coherent melodic lines. Thus, Baroque chord progressions are usually quite consonant, and twelve-tone chord progressions quite dissonant, on all four counts. Other combinations of the above factors are common: some impressionist styles have consonant chords but dissonant chord relationships and weak melodic and tonal structure; and some jazz styles have dissonant chords, but consonant chord relationships and strong melodic and tonal structure.

Chord progressions without passing notes are *homorhythmic*: all parts move in synchrony. In *polyphonic* (or contrapuntal) music, by contrast, the parts move in independent rhythms. The model to be developed in Chap. 4 concerns the perception of simultaneities and of pitch relationship between simultaneities, i.e. of homorhythmic music.

Ideally, perceptual grouping (Sect. 2.3.3) in homorhythmic music is entirely vertical – tone sensations are grouped only because they are close in time. In polyphonic music, perceptual grouping is ideally only horizontal, due to grouping in pitch and timbre (streaming). In reality, perceptual grouping of both kinds occurs in both styles, but vertical grouping is more pronounced in the case of homorhythmic music, and horizontal grouping is more pronounced in polyphonic music. The present study is mainly concerned with homorhythmic music, in which vertical grouping is assumed to predominate.

The term "chord progression" in this study refers to a limited number of chords, perceived as a group – in a similar way that the tones of a melodic phrase are grouped. All parts of a chord progression are assumed to be simultaneously accessible to the attention of the listener just after the progression has been completed. The duration of a chord progression is therefore limited to the maximum duration of auditory sensory memory, i.e. 2–10 s (Sect. 2.2.3). If the chords of the progression follow each other at a maximum rate of about $4\,s^{-1}$ (the rate at which consonants are typically produced in speech), and the duration of short-term memory is 5 s, then the maximum number of chords a progression may contain is $4 \times 5 = 20$. The progressions analyzed in this study (Sect. 6.3.1) comprise no more than ten chords and may therefore be assumed to be perceptible as wholes.

4. Model

The quantitative relationship between physical and experiential properties of tones, chords and progressions is simulated by means of a model. The input to the model comprises either the amplitude spectra or the musical notes of each tone or chord. Evaluation of masking effects leads to an estimate of the audibility of each pure tone component. The perception of complex tones is simulated by matching the pitches of audible pure tone components against those of a template, representing the audible components of a typical complex tone. Calculated audibilities of pure and complex tone components are used to estimate the "tonalness" and "multiplicity" of a simultaneity, and the salience of each tone sensation. Calculated salience values enable the evaluation of pitch commonality and pitch distance between sequential sounds. These are combined to model the results of pitch analysis and tone/chord similarity experiments.

4.1 General Aspects

4.1.1 Aim, Form and Implementation

The prediction of the pitch properties of a tone simultaneity on the basis of its waveform or frequency spectrum entails extensive understanding of psychoacoustical data and theory [Terhardt et al. 1982b]. Similarly, the determination of the musical function of a chord from the notes of a musical score requires considerable familiarity with music and music theory.

The model described in this chapter aims to facilitate the prediction and understanding of these relationships. It formalizes experimental data and psychoacoustical theory on the perception of tone simultaneities in such a way that: the data and theory are logically, concisely and conveniently expressed; the theory may be quantitatively tested (Chap. 5); and the theory may be applied in music theory, analysis and composition (Chap. 6).

Mathematics itself cannot *explain* music, it can only *describe* it [cf. De la Motte-Haber 1985]. In the model, mathematics is used to describe psychoacoustical data and theory concerning the perception of tone simultaneities. The mathematics is drawn together into a single *algorithm*: a mathematical procedure which produces a unique output for a given set of input parameters.

Existing models of music theory and analysis such as those by Alphonce [1980], Rahn [1980] and Smoliar [1980] are based primarily on the experience and intuition of music theorists such as Schenker [1906, 1935]. The present model is similar in that it is also largely based on experience and intuition, in this case of psychoacousticians such as Terhardt. It has the advantage that it draws upon a higher proportion of experimental data and scientifically testable theory than its music-theoretical relatives, and the disadvantage that it fails to account for cultural or cognitive aspects of music perception such as the perceptual organization of relatively long time spans (compare Schenker's *middleground* and *background*).

The model may be regarded simply as a *black box*. Of primary interest is the relationship between the input and the output [cf. West et al. 1987]. When suitably implemented in the form of a computer program, the model may be used without prior knowledge of how and why it works. Its use may be justified solely in terms of the success with which it models the results of experiments and the conventions of music theory.

A model such as this would be impractical without modern computer technology. In this study, computers were used to perform the long strings of calculations required by the model, to automate experiments by which the model was tested, to analyze and reduce experimental data, and to compare calculations of the model with experimental results so as to allow objective judgment of the model's predictive power. By means of computers, various versions of the model could be quickly compared to see how well they modelled the results of particular experiments. Inferior versions and versions which did not improve modelling performance despite increased complexity and computation time could then be discarded. Finally, computers enabled the model to be applied in music theory and analysis (Chap. 6).

4.1.2 Formulation and Assessment

A variety of mathematical operations and functions were drawn upon to create the model. Among the more commonly used functions were logarithms and exponentials, functions which have traditionally played important roles in psychophysical modelling [Fechner 1860; Stevens 1957]. Another common operation in the model is summation (e.g. summations of functions of pitch over the range of hearing). Summations are used to simulate relationships between relatively analytical and relatively holistic sensations (Sect. 2.3.3). Mathematical forms were regarded as appropriate if they logically and non-arbitrarily embodied appropriate theoretical concepts, and so allowed theoretical ideas (e.g. concerning the supposed sensory basis of Western harmony) to be tested as objectively as possible. A further criterion was efficiency: mathematical forms were preferred if they were relatively simple to understand and to calculate, and enabled the results of appropriate experiments to be modelled within a reasonable margin of error.

The model, as it stands, is neither perfect nor final. In its present state it is sufficiently logical, simple and accurate for a range of music-theoretical applications. It could, however, be changed in innumerable ways to satisfy future requirements, or as new theoretical possibilities emerge.

The model was tested against the results of psychoacoustical experiments (Chap. 5) based on subjective ratings of, and comparisons between, tones and chords. Responses were influenced by interindividual variations (such as listener's personalities and abilities, how well they were able to concentrate at the time of the experiment, etc.) and the context of each experimental trial. The model accounts for certain interindividual variations and contextual effects by means of four *free parameters.* Each parameter is supposed to reflect *how analytically sound is perceived* by the listener at a certain level. The first level concerns the analysis of a simultaneity into pure tone components: the more analytically one listens, the more clearly pure tone components can be resolved, and so the greater is the number of audible pure tone components, see eq. (4.14) below [E. J. Gibson 1953]. The second level concerns the perception of individual tones within a simultaneity: the more analytically one listens, the more likely one is to notice pure tone sensations by comparison with complex tone sensations (4.19) [Terhardt 1972, 1974a]. The third level concerns the simultaneous perception of tones: the more analytically one listens to a simultaneity (such as a musical chord), the more tones one notices at the one time (4.24). The fourth level concerns perceived relationships between sequential sounds: the more analytically such a relationship is heard, the more it is supposed to be influenced by pitch commonality as opposed to pitch proximity (4.34, 37).

4.1.3 Culture-Specific Aspects

Listeners' musical experience – in particular, their experience of sounds similar to those heard in the experiments – has a considerable effect on experimental results. This was not accounted for at all in the model. Instead, specific cultural conditioning effects were explored by looking at systematic discrepancies between calculations and averaged experimental responses. The model is nonetheless culture-specific in several respects. The aim of the model is primarily to explain aspects of the perception of Western chord progressions. Consequently, it is formulated in such a way that it can only be applied to the theory and analysis of Western music.

The following list gives some idea of the extent of the model's ethnocentricity.

i) Pitches and frequencies are limited to the chromatic scale throughout the model. This limits application of the model to musical cultures whose scales may adequately be represented as subsets of the chromatic scale (Sect. 3.3.2). Note that exact tuning of the chromatic scale is not specified in the model.

ii) Emphasis is placed on how analytically sounds are heard (the four free parameters). This may reflect a kind of Western analytical bias.
iii) The importance of pitch for musical structure is emphasized. The relative importance of pitch (as opposed to rhythm, timbre, dynamics, etc.) differs across musical cultures.
iv) The importance of sounds with harmonic spectra is emphasized, and perceptual conditioning by sounds with nonharmonic spectra (important in many world musics) is ignored.
v) The importance of consonance – and of tonalness, pitch commonality and pitch distance as sensory components of consonance – is emphasized.

The psychoacoustics of hearing has many aspects, some of which are well understood, some yet to be explored. There presumably exist many universal properties of human hearing which are not exploited by Western music, but which influence non-Western musics in important ways and may therefore be described as sensory bases of these musics. The identification and investigation of such properties would require intimate knowledge of the music(s) in question, and is beyond the present scope.

4.1.4 Comparison with Terhardt's Model

The present model is based on and inspired by the *Algorithm for the extraction of pitch and pitch salience from complex tonal signals* of Terhardt, Stoll and Seewann [1982b], hereafter referred to as "Terhardt's model". The present model differs from Terhardt's in several important respects.

Terhardt's model operates directly on an amplitude spectrum, created by the fast Fourier transform (FFT) of a segment of the waveform of a sound, and includes a procedure for the extraction of tonal components from the spectrum. In the cases of interest for the current study, spectral analysis was unnecessary. In experimental testing of the model (Chap. 5), the spectra from which the sounds were synthesized were input directly to the model. In music theoretical applications of the model, the spectra of individual complex tones are specified by the model, so that only note names need to be supplied.

Terhardt's model is primarily intended to predict the *exact pitch* of the most prominent tone sensation in a simultaneity. It therefore deals with *pitch shifts* in considerable detail. Music theory is primarily concerned with relationships between pitch *categories*, i.e. notes (Sect. 2.5.3). It turns out that musical pitch relationships may be modelled without reference to pitch shifts. Consequently, pitch shifts are neglected altogether in the present model.

Tone saliences (Terhardt: *pitch weights*) are expressed as absolute values in the new model. Tone salience is defined as the probability of noticing a tone (i.e. the probability of experiencing the corresponding tone sensation). In Terhardt's model, pitch weights are used primarily to determine which pitch (i.e. which tone sensation) in a simultaneity is most prominent; they are not regarded as absolute values, nor are they given a definite psychoacoustical meaning.

Finally, the new model proceeds further in the direction of music theory than Terhardt's, by estimating the strength of the pitch relationship (pitch commonality, pitch proximity) between sequential musical tones or chords.

Because the two models are so different, the degree to which they may usefully be compared is limited. Ultimately, the models should be judged separately on their individual merits: how logically they embody the theories on which they are based, how simple they are, and how closely they predict experimental results corresponding to their separate aims.

4.2 Input

4.2.1 Pitch Category

Following a notation adopted by the American Standards Association [1960], the *register* of a note is written as a subscript to the note's letter name. For example, middle C is called C_4. Each register runs from C up to B, so the B next to middle C is B_3. The lowest pitch register (corresponding roughly to the range 16–32 Hz) is "register 0"; the highest (roughly 8–16 kHz), "register 9". The modern piano keyboard includes 7 complete registers (1–7).

The *chroma* (or pitch class) of a note is defined in the model as its distance in semitones from the nearest C below. So the chroma of any C is 0, of any G is 7, and so on. No distinction is made between enharmonically equivalent note names when they are expressed as chromas. For example, E sharp, F natural and G double flat all lie 5 semitones above C, and so have chroma 5.

The *pitch category* of a musical note in the model is obtained by multiplying its register by twelve and adding its chroma. For example, the pitch category of the note A_4 (440 Hz) is $12\times4+9=57$. The range of the modern piano (A_0 to C_8) may be expressed as a pitch category range of 9–96.

4.2.2 Experiments

The sounds presented in the modelled experiments (Sects. 5.2, 5.3, 5.6, 5.7) were initially specified by the pitch categories (P) and sound pressure levels (SPL) of their tone components. When realizing the sounds, the frequency levels FL [Fletcher 1934; Young 1939] of tone components were set equal to their pitch categories P, i.e. tone components were tuned to equal temperament:

$$\mathrm{FL}(P) = P \ . \tag{4.1}$$

It was unnecessary to account for octave stretch (Sect. 3.3.3), as sounds in each experimental trial covered relatively small frequency ranges.

In the experiments on the similarity of synthetic tones (Sects. 5.5, 5.6), fundamentals were tuned according to (4.1), and overtones were exact multiples of fundamental frequencies. In other experiments using synthetic sounds (Sects. 5.2, 5.3 and 5.7), (4.1) was applied to each pure tone component individually, so that spectra were entirely equally tempered.

When synthesizing sounds for the experiments, frequencies f [Hz] of pure tone components were obtained from their frequency levels FL by the equation:

$$f(\mathrm{FL}) = 2^{(\mathrm{FL}-57)/12} \times 440 \ . \tag{4.2}$$

Similarly, the pressure amplitude p of each pure tone component was calculated from its specified SPL by

$$p(\mathrm{SPL}) = k \times 10^{\mathrm{SPL}/20} \ , \tag{4.3}$$

where the factor k depends on sound amplification.

The frequency spectrum of each sound heard in the modelled experiments was input to the model as an array of real numbers in the form SPL(P), where pitch category P ranged from 19 to 96. Empty pitch categories were assigned large negative SPLs (i.e. zero amplitudes).

4.2.3 Auditory Level

The threshold of audibility in free field for pure tones is expressed in the model as a *threshold level* (TL), in dB (SPL), as a function of pitch category P:

$$\mathrm{TL}(P) = 91 - 49 \log_{10}(P-7) \ . \tag{4.4}$$

Over the pitch range of the modern piano (P from 9 to 96), threshold level values calculated in this way differ by 1 dB (average) to 2 dB (maximum) from values calculated by Terhardt's [1979a] formula for threshold level. Terhardt's formula had been fitted to data averaged over a number of people with good hearing [Zwicker and Feldtkeller 1967] over the entire audible range. The two formulas are compared in Fig. 4.1. Beyond $P = 96$ (4.2 kHz), the two formulations diverge rapidly and (4.4) becomes invalid. Frequencies in this range were not used in the experiments of this study.

The *auditory level* (YL) of a pure tone component is defined here as its level in decibels above the threshold of audibility:

$$\mathrm{YL}(P) = \max\{\mathrm{SPL}(P) - \mathrm{TL}(P), 0\} \ , \tag{4.5}$$

where the operator "max" selects the maximum of the two values in curly brackets.

A pure tone component with positive auditory level YL may be inaudible due to masking by other simultaneous components. In this case the component is said to have positive auditory level YL but zero *audible* level AL, see (4.16). Both YL and AL are supposed to apply to an idealized average listener with good hearing.

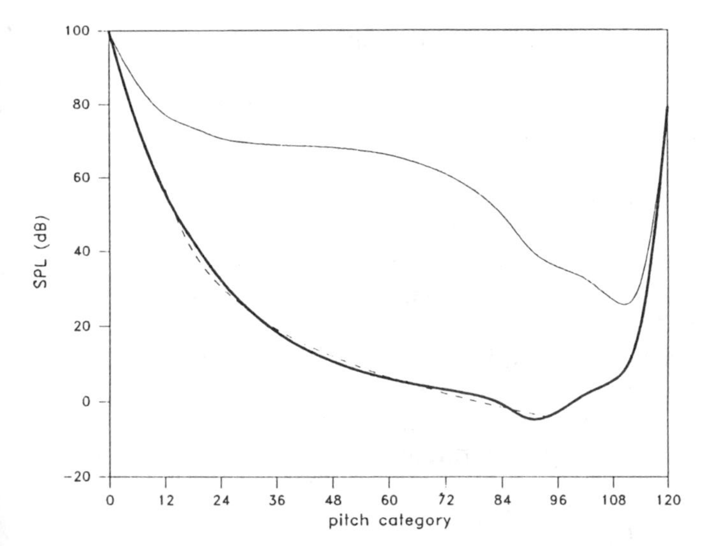

Fig. 4.1. Specified sound levels of musical tones. *Bold line*: threshold of audibility according to Terhardt [1979a]. *Broken line*: threshold of audibility according to (4.4). *Upper curve*: specified SPLs of fundamentals of musical tones according to (4.8), relative to the bold line

4.2.4 Applications

In most applications of the model (Chap. 6), the user inputs only the type and pitch category of each tone (component). Tone type may be "pure", "full complex" or "octave-spaced". In each case, the model specifies the auditory levels of pure tone components in a way which is consistent with the amplitude spectra of typical musical sounds.

The pure tone option is useful in the case of non-harmonic complex sounds. Only the pitch categories P of pure tone components are supplied by the user. Each component is then assigned an auditory level yl as follows:

$$\mathrm{yl}(P) = \frac{P(120-P)}{60} . \tag{4.6}$$

According to this function (which is just an arc of a circle), assigned auditory levels of pure tone components approach 60 dB over the central pitch range (roughly corresponding to the spectral dominance region [Fletcher and Galt 1950]) and fall to zero at the upper and lower thresholds of pitch.

Normally, musical sounds are made up of complex tones, specified by musical note names. So for music-theoretical or music-analytical work it is convenient for the model to accept musical note names as input. In the present model, each note is assumed to be realized as a full complex tone with all harmonics from the first to the sixteenth. Harmonics above the tenth are included even though they are rarely audible [Plomp 1964; Terhardt 1979a] as they can still contribute to the masking of other components, and thereby to the overall sound.

If the fundamental (first harmonic) of a musical tone belongs to pitch category P_1, then the nth harmonic belongs to category P_n, as follows:

$$P_n = P_1 + \text{int}\ \{12 \log_2 (n) + 0.5\}\ . \tag{4.7}$$

In this equation, the operator "int" (integer part) converts the frequency intervals between the components of a harmonic series into whole numbers of semitones, and the "0.5" ensures that values are correctly rounded. The values of $(P_n - P_1)$ generated by this equation, i.e. pitch distances above the fundamental in semitones for $n = 1$ to 16, are 0, 12, 19, 24, 28, 31, 34, 36, 38, 40, 42, 43, 45, 46, 47 and 48: the harmonic series in whole semitones (Fig. 1.1).

The auditory level yl of the nth harmonic is specified in the model by

$$\text{yl}(n) = \frac{P_n(120 - P_n)}{60} \left(1 - \frac{(P_n - P_1)}{120}\right) . \tag{4.8}$$

This expression satisfies the following requirements:

- On average, SPLs of harmonics of musical tones exceed threshold level by around 50 dB.
- Auditory levels fall to zero at the upper and lower thresholds of pitch ($P_n = 0$ and 120).
- Auditory levels are normally low in the lowest two pitch registers (below C_2, $P = 24$) due to the high threshold of audibility in these registers (Fig. 4.1).
- Pure tone components are seldom important for pitch perception in the highest two registers (above C_8, $P = 96$) due to spectral dominance (Sect. 6.1.2).
- The levels of coinciding pure tone components belonging to different complex tones in well-balanced musical chords are normally such that a component of higher harmonic number has a lower level than a coincident harmonic of lower harmonic number. For example, the 3rd harmonic of C_4 ($P_3 = 48 + 19 = 67$) is specified by (4.8) to be about 10 dB lower in level than the 1st harmonic of G_5 ($P_1 = 67$).
- Average SPL gradients for the harmonics of complex tones are shallower for lower tones (−8 to −10 dB per octave for fundamentals in the range C_1 to C_3) and steeper for higher tones (−12 to −17 dB per octave for C_4 to C_6) according to this equation [Parncutt 1987a].

In music-theoretical applications of the model, it is sometimes useful to analyze the pitch properties of chords composed of octave-spaced tones (Sects. 6.1.6, 6.2.1, 6.2.3–5). An octave-spaced tone at chroma c has pure tone components in every register $r = 0$ to 9 with pitch categories P_r given by

$$P_r = c + 12r \ . \tag{4.9}$$

In the model, these components are assigned auditory levels yl according to

$$\mathrm{yl}(r) = \frac{P_r(120 - P_r)}{60} \ . \tag{4.10}$$

In musical chords, pure tone components which fall into the same pitch category P are *incoherent*, i.e. their phase relationship is random. The auditory levels $\mathrm{YL}(P)$ of such components are combined according to elementary acoustics by adding their effective intensities (which are proportional to the squares of their sound pressures) by the formula

$$\mathrm{YL}(P) = 10 \log_{10} \sum 10^{\mathrm{yl}(P)/10} \ , \tag{4.11}$$

where $\sum$ denotes summation over all components falling in the same pitch category P, and the $\mathrm{yl}(P)$ are the individual auditory levels of each of the coinciding pure tone components.

4.3 Masking and Audibility

4.3.1 Critical Bandwidth

Fletcher [1940] suggested that the audible frequency range is divided into separate *critical bands*, as if the auditory system contained filters with variable centre frequencies. The loudness of a band of noise of constant power is constant for bandwidths less than a critical bandwidth, and increases for larger bandwidths [Zwicker et al. 1957]. Critical bandwidth, according to this definition, equals about 3 semitones above 500 Hz (C_5), and approaches a linear relationship with frequency at lower frequencies. Listener's responses in a variety of other experiments also change abruptly as bandwidth is increased beyond one critical bandwidth [Zwicker 1961; Scharf 1970].

There is some disagreement in the literature on the size of critical bandwidth at low frequencies. Zwicker et al. [1957] found that critical bandwidth below 500 Hz is approximately constant at 100 Hz [see also Zwicker and Terhardt 1980]. More recent measurements by other authors, using various methods, generally yielded lower estimates of critical bandwidth at low frequencies. Some published results are set out in Table 4.1.

In the present model, critical bandwidth W_{cb} is expressed in semitones as a function of the pitch category P at the centre of a band as

$$W_{\mathrm{cb}}(P) = \frac{5}{1 + x/\sqrt{x^2 + 44}} \ , \quad \text{where} \quad x = P/5 - 10 \ . \tag{4.12}$$

Table 4.1. Critical bandwidth [Hz] at low frequencies

Centre frequency [Hz]	125	200	250	500
Zwicker et al. [1957]	100	100	100	110
Patterson [1976]				80
Houtgast [1977]			50	80
Fidell et al. [1983]	40	50		
Calculated values (text)	60	70	80	110

This formulation was obtained by comparing calculations based on (4.13) below with experimental data (Table 4.1).

In the simulation of masking between pure tone components, it is useful to transform their frequencies or pitch categories onto a scale in which equal distances correspond to equal numbers of critical bandwidths. The accepted name for such a scale, *critical-band rate* [Zwicker 1961], is confusing because it implies that the scale is found by differentiation. In fact it is found by *integration* of the reciprocal of critical bandwidth. This conceptual difficulty may be avoided by anglicizing Zwicker's term *Tonheit* (literally: "tone-ness"; translated "tonalness" by Scharf [1970]) and renaming the critical-band rate scale *pure tone height.* The new name makes it clear that the scale is applicable only to the (spectral) pitch of pure tones and pure tone components – not to the (virtual) pitch of complex tones, such as in music. "Height" is used in

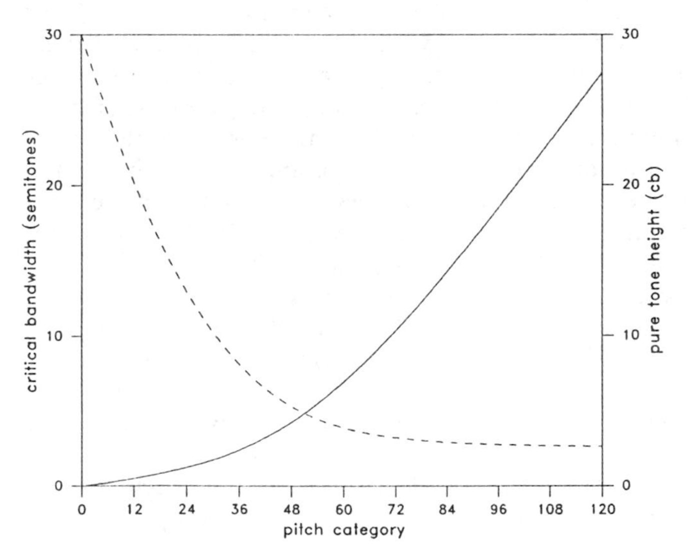

Fig. 4.2. Critical bandwidth and pure tone height. *Broken line, left scale*: critical bandwidth in semitones according to (4.12). *Full line, right scale*: pure tone height in critical bandwidths [cb] according to (4.13)

preference to "pitch", as "pitch" has musical connotations. The pure tone height scale is equivalent to the mel scale of pitch [Zwicker and Feldtkeller 1967].

Pure tone height H_p in critical bands [cb] may be derived from (4.12) by integrating its reciprocal and setting $H_p = 0$ at $P = 0$. The result is

$$H_p(P) = \sqrt{x^2+44}+x-2 \ , \quad \text{where} \quad x = P/5-10 \ . \tag{4.13}$$

Equations (4.12) and (4.13) are graphed in Fig. 4.2.

4.3.2 Masking

Masking may be described as the mutual drowning out (or inhibition, see [Moore 1982]) of one sound by another. Every pure tone component in a moderately loud sound partially masks every other component lying within a maximum distance of about 3 critical bands [Terhardt et al. 1982b]. The degree to which a component of auditory level YL in pitch category P' masks a component in a different category P is expressed in the model in terms of the effective reduction (in dB) of the audible level of P due P', as follows:

$$\mathrm{ml}(P,P') = \mathrm{YL}(P') - k_M \, |H_p(P') - H_p(P)| \ , \tag{4.14}$$

where ml stands for masking level and the expression within vertical bars (absolute value symbols) is the distance between P and P' in critical bandwidths. The "masking parameter" k_M is the first free parameter of the model. It has a typical value of about 25 dB/cb (Sect. 5.8.1). Note that ml in (4.14) may be negative; this is rectified in (4.15) below.

According to (4.14), the masking pattern (or audiogram) of a pure tone, when depicted as a graph of SPL against pure tone height, is triangular, with equal upper and lower gradients of k_M dB/cb (Sect. 2.2.2). In fact, the upper and lower gradients of masking patterns generally differ, and depend on level and frequency [Terhardt 1979a]. The masking parameter's optimal value of 25 mostly exceeds values used by Terhardt [1979a] to calculate masking, which are centred in the range 15–25 dB/cb. A possible explanation is that slightly subthreshold pure tone sensations can contribute to the formation of complex tone sensations. This would be consistent with the finding of Moore and Rosen [1979] that a residue tone can evoke a pitch at the missing fundamental even if its harmonics are (apparently) completely masked.

The interaction of several maskers has been investigated extensively [Zwicker and Herla 1975; Lutfi 1985; Moore 1985], but no rule has been found by which the effects of simultaneous maskers may be combined in a way which is accurate, yet appropriately simple for music-theoretical purposes. Terhardt's [1979a] solution was to combine contributions to the masking of a particular tone component by more than one other component by adding equivalent sound pressure *amplitudes.* This method is relatively simple, and makes plausible predictions of the number of audible harmonics in typical complex tones [Terhardt 1979a]. It is therefore retained in the present model:

$$\mathrm{ML}(P) = \max\left\{20 \log_{10}\left(\sum_{P' \neq P} 10^{\mathrm{ml}(P,P')/20}\right), 0\right\} , \tag{4.15}$$

where the summation is carried out over all values of P' not equal to P, and the "max" function prevents ML from becoming negative (in the case of no maskers). This formation corresponds to that proposed by Lufti [1985] for the case of "appreciably overlapping maskers" (such as the pure tone components of full complex tones and musical chords).

The masking algorithm described in this section is only an approximation to that described by Terhardt [1979a], which itself will become obsolete as the Fourier-t-transform with appropriate frequency and amplitude dependencies is introduced to account for auditory masking [Terhardt 1985]. However, it is accurate enough for music-theoretical work, in which rough estimates of masking levels are sufficient (Sect. 6.1.1).

4.3.3 Audibility

The *audible level* (AL) of a pure tone component (Terhardt: *SPL excess*) is defined as its level above masked threshold. It is calculated in the present model by subtracting the masking level ML at the pitch category P of the component from the component's auditory level YL:

$$\mathrm{AL}(P) = \max\{\mathrm{YL}(P) - \mathrm{ML}(P), 0\} . \tag{4.16}$$

According to Hesse [1985], the perceptual prominence (*Ausgeprägtheit*) of the pitch of a partially masked pure tone relative to that of a clearly audible reference tone is equal to about 1/20 of its audible level in decibels at low audible levels, and approaches 1 (saturates) at high audible levels. Terhardt et al. [1982b] defined *spectral pitch weight* in a way which closely fits Hesse's data on the pitch prominence of pure tones. They allowed the audibility of each pure tone component to saturate with increasing audible level as follows:

$$A_{\mathrm{p}}(P) = 1 - \exp\{-\mathrm{AL}(P)/15\} , \tag{4.17}$$

where A stands for audibility, and the subscript p stands for pure tone. The saturation of audibility with increasing level has the effect that the tonalness (Sect. 4.4.3) of a clearly audible simultaneity is roughly independent of level.

4.4 Recognition of Harmonic Pitch Patterns

4.4.1 Harmonic Template

Complex tone perception may be regarded as a direct process: complex tone sensations are merely experiences accompanying the perception of tone sources [Gibson 1966] (Sect. 2.1.3). This state of affairs is not reflected in a psychoacoustical model, in which analytical sensations are regarded as basic. In the present model, the perception of complex tones is simulated as if it were mediated by the formation of pure tone sensations (Sect. 2.4.3).

According to the theory of virtual pitch [Terhardt 1972, 1974a] complex tone sensations result from the spontaneous recognition of (normally unnoticed) harmonic patterns among the (spectral) pitches of pure tone sensations. The present model stimulates the relationship between pure and complex tone sensations by searching for harmonic pitch patterns among pure tone sensations. A harmonic template, whose components form an idealized harmonic pattern, is used to estimate the importance of each pattern found (Sect. 2.4.3). The form of the template resembles the pitch pattern of the audible harmonics of a typical harmonic complex tone (Fig. 4.3).

The pure tone sensations contributing to the formation of a complex tone sensation at pitch category P_1 (the 1 denotes first harmonic) fall in pitch categories P_n, $n = 1, \ldots, 10$, as specified in (4.7). The degree to which pure tone sensations contribute to the formation of complex tone sensations becomes progressively less for higher harmonic numbers n [Ritsma 1967]. This is accounted for by assigning a weight (W) to each element of the template:

$$W_n = 1/n \ . \qquad (4.18)$$

These weights resemble the audibilities of harmonics of typical complex tones, such as speech vowels [Terhardt 1979a]. Normally, no more than ten components are audible. The template is therefore limited to ten components, i.e. $n = 1, \ldots, 10$ in both (4.7) and (4.18). This limitation has the advantage that no decision has to be made about the pitch category of the eleventh component, which lies 41.5 semitones above the fundamental, on the borderline between two pitch categories. Note that the first component of the template has twice the weight of the second, reflecting the large difference between the pitch saliences of full complex tones and residue tones [Stoll 1983].

4.4.2 Complex Tone Sensations

The formation of complex tone sensations is simulated by shifting the harmonic template (Fig. 4.3) through the musical pitch range in steps of one semitone. At each step, pitch matches are sought between the components of the template and the pure tone sensations of the sound. Whenever one or more

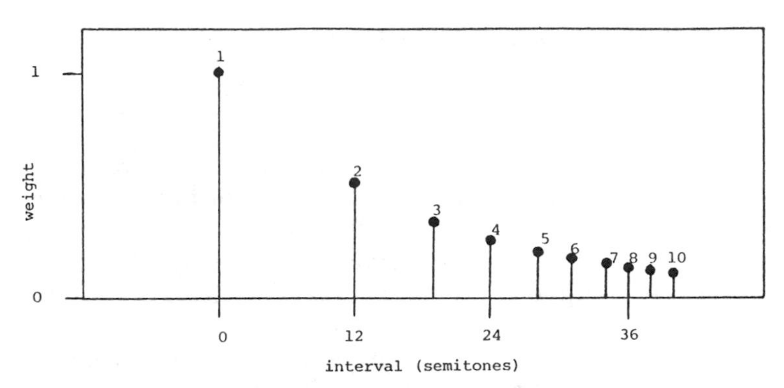

Fig. 4.3. Template for the simulation of harmonic pitch pattern recognition. Intervals are given by (4.7), weights by (4.18)

matches are found, a complex tone sensation is generated, whose pitch category (P_1) is that of the template's lowest component, and whose audibility (A_c) depends on the numbers and audibilities (A_p) of pure tone components matching template components as follows:

$$A_c(P_1) = \left(\sum_n \sqrt{W_n A_p(P_n)} \right)^2 / k_T \ , \tag{4.19}$$

where P_n is given by (4.7) and W_n by (4.18).

The "tone perception parameter" k_T is the second free parameter of the model. It is supposed to indicate how analytically tones are perceived, depending on listener and context. If k_T is low (about 1), the pure tone components of a complex sound are much less audible than complex tone components, and so are noticed only rarely. If k_T is high (about 10), pure and complex tone components compete with each other for the listener's attention on roughly equal terms. The value $k_T = 3$ was found to be typical in the experiments of this study (Chap. 5) and so was adopted for music-theoretical applications (Chap. 6).

Equation (4.19) may be regarded as a mathematical formulation of Houtgast's [1976] statement that "the potential of a multicomponent signal in evoking a particular low pitch [complex tone sensation] can be understood as a simple combination of the potentials of the individual components in evoking a particular subharmonic pitch" (p. 409). The "simple combination" chosen in the current model is not a simple sum but the square of a sum of square roots. In other words, complex tone audibility A_c is assumed to be proportional to the *square* of the number of template matches with audible pure tone components. This ensures that the calculated saliences of the subharmonic tone sensations of a pure tone are appropriately weak (Sects. 2.4.5, 6.1.5). Note that the value of A_c does not increase exponentially as harmonics are added to a complex tone, because mutual masking of harmonics causes their audibilities A_p to decrease markedly.

In Terhardt's model, contributions to the salience of a virtual pitch are added (linearly) from *pairs* of spectral pitches. Consequently, pure tones are assigned single, unambiguous pitches. Equation (4.19) was chosen here in favour of Terhardt's formulation, as it is easier to calculate (requiring one loop instead of two nested loops), and it simulates the concept of template matching in a more straightforward manner. Also, spectral and virtual pitch weight are defined in Terhardt's model to depend on the absolute frequency of tone components (spectral dominance). Various versions of these dependencies were tried out in in (4.17, 19), but none improved the correlation between calculations and the results of the experiments reported in Chap. 5. Some consequences of neglecting spectral dominance in the model are investigated in Sect. 6.1.2.

Isolated pure tones of moderate loudness in the middle pitch range have audible levels of about 60 dB. According to (4.17), such tones have audibilities approaching one. Isolated complex tones of moderate loudness in the middle pitch range, according to the model [with $k_T = 3$ in (4.19)], have audibilities five to six times greater than this. (The same applies for spectral and virtual pitch weights in Terhardt's model.) This difference is understandable in terms of the number of sources of information about the pitch of a tone – pure tones have only one such source, while complex tones have several (the harmonics) – and in terms of the relative importances of pure and complex tones in the auditory environment.

In analytical listening, it is possible to switch attention from a complex tone as a whole to the pure tone component at its fundamental [Schouten 1940]. But is is not possible to attend to both at once, as their pitches are the same (or very close). Following Terhardt's model, this implies that the overall audibility $A(P)$ of a tone component (i.e. audibility regardless of whether that component is pure of complex) is given by the maximum of the pure and complex tone audibilities in that pitch category:

$$A(P) = \max\{A_p(P), A_c(P)\} \ . \tag{4.20}$$

4.4.3 Tonalness

The *pure tonalness* of a simultaneity is defined here to depend on the number and audibilities of pure tone components (Sect. 3.2.2). Following Aures [1984], pure tonalness may be modelled by quadratic addition of the audibilities of pure tone components:

$$T_p = \sqrt{\frac{\sum_P A_p(P)^2}{5.2}} \ . \tag{4.21}$$

The factor 5.2 scales pure tonalness values so that the calculated pure tonalness of a complex tone at middle C [specified by (4.7, 8)] is 1. Examples of calculated pure tonalness values are given in Sect. 6.1.4.

The *complex tonalness* of a simultaneity is assumed here to be proportional to the maximum virtual pitch weight in Terhardt's model. In the terminology of the present study, it is proportional to the audibility of a simultaneity's most audible complex tone component (Sect. 3.2.2):

$$T_c = \max_P\{A_c(P)\}/6.2 \ . \tag{4.22}$$

As before, the factor 6.2 scales complex tonalness values so that the calculated complex tonalness of a complex tone at middle C [specified by (4.7, 8)] is 1.

4.5 Salience

4.5.1 Multiplicity

In this study, the number of tones simultaneously noticed in a sound (e.g. in a musical chord) is called its *multiplicity* (Sect. 2.3.4). Multiplicity is assumed in the model to depend partly on a sound's pitch configuration – its configuration of tone audibility as a function of pitch category – and partly on how analytically the sound is perceived.

An initial, unscaled estimate M' of the number of tones noticed simultaneously in a sound may be made by assuming that the sound's most audible (pure or complex) tone component is noticed with a probability of 100%, while other, less audible tone components are noticed with probabilities proportional to their calculated audibilities:

$$M' = \frac{\sum_P A(P)}{A_{\max}} \ , \tag{4.23}$$

where $A_{\max}$ is the maximum audibility in the sound, i.e. the audibility of the most audible (pure or complex) tone component.

When perceived holistically, a tonal sound evokes a single tone sensation. For example, a holistically perceived chord built from octave-spaced tones evokes a single tone sensation which usually corresponds to its root (Sect. 6.1.6). In this case, the actual number of simultaneously noticed tones M equals one, regardless of the value of M' in (4.23). A *power law* relationship [Stevens 1957] between M and M' allows for this: any value of M' may be scaled to a value of $M = 1$ by raising it to the power zero. In general, M' may be appropriately scaled by raising it to some power k_S:

$$M = (M')^{k_S} \ . \tag{4.24}$$

The "simultaneity perception parameter" k_S is the third free parameter of the model. It may take any positive or zero value; the higher k_S, the higher M, and the more analytically simultaneities are perceived. Calculations according to (4.23, 24) fitted the results of the multiplicity experiment (Sect. 5.2.3) most closely when the parameter was set to a value of about 0.5. So M is set equal to the square root of M' in music-theoretical applications of the model (Chap. 6).

Equations (4.23, 24) disregard the important effects of relative onset times of the notes of a chord [Rasch 1978] and coordination of component amplitude envelopes [McAdams 1984] on multiplicity. The auditory system is remarkably sensitive to these subtle effects, using them to establish the actual number of sound sources contributing to a particular sound (e.g. a musical chord played by several instruments). These cues were absent from the sounds presented in the multiplicity experiment (Sect. 5.2.2), allowing the formulation presented here to be tested relatively directly. In performed music, where onsets and amplitude envelopes are asynchronous and amplitude envelopes independent, the number of simultaneously perceived tones corresponds more closely to the actual number of simultaneous notes, provided this number remains small (say, no more than four).

When sounds (such as musical tones or chords) are perceived holistically (i.e. when they are assigned only one pitch at a time), the variable M may be interpreted as a measure of *pitch ambiguity*: an estimate of the number of different pitches a sound *could* have, where each pitch is weighted according to the salience of the corresponding tone sensation. The greater the pitch ambiguity of a sound, the greater the number of different pitches which could be assigned to the sound when it is perceived on different occasions and in different contexts.

4.5.2 Tone Salience

The salience (S) of an individual tone component is defined as its probability of being noticed. Assuming independent probabilities, and given that the tone simultaneity itself has been noticed, it follows that the sum of all the tone saliences $S(P)$ in a sound equals the number of simultaneously noticed tones M. In addition, tone salience is assumed to be proportional to tone audibility A. The following expression satisfies these criteria:

$$S(P) = \frac{A(P)}{A_{\max}} \frac{M}{M'} . \tag{4.25}$$

Assuming that only one tone is perceived at a given time in each pitch category of the chromatic scale, the variable S may also be interpreted as the "salience of a pitch category".

In music-theoretical applications, tone salience may be expressed in terms only of the audibilities A of other simultaneous tone sensations, by substituting $k_S = 0.5$ into (4.24) and combining (4.23–25):

$$S(P) = \frac{A(P)}{\sqrt{A_{\max} \sum_P A(P)}} \; . \tag{4.26}$$

4.5.3 Chroma Salience

In music-theoretical applications such as the determination of the root of a chord (Sect. 6.1.6) and the tonality of a progression (Sect. 6.2.3–5), it is useful to evaluate the saliences of the twelve chroma (pitch classes) in the chromatic scale. Two possible measures of chroma salience are considered here: *chroma tally* and *chroma probability.*

Chroma tally is defined as the average number of times a chroma is noticed in a musical element or passage. Assuming that tone saliences are independent probabilities, chroma tally $S_t(c)$ is given by

$$S_t(c) = \sum_r \sum_s S(c+12r) \; , \tag{4.27}$$

where the summations are carried out over all pitch registers r and simultaneities s in a homorhythmic (tone or chord) progression.

Chroma probability $S_p(c)$ is defined as the probability that a chroma is noticed (at least once) in a musical element or passage. Again assuming independent tone saliences,

$$S_p(c) = 1 - \prod_r \prod_s \left[1 - S(c+12r)\right] \; , \tag{4.28}$$

where $\prod$ denotes a product over all pitch registers r and simultaneities s in a homorhythmic progression.

4.6 Sequential Pitch Relationship

4.6.1 Pitch Commonality

The *pitch commonality* of a pair of sounds is assumed to be proportional to the number of pitches the sounds have in common (Sect. 3.2.3), i.e. the number of pitch categories containing noticed tones in both sounds. In the special case that the sounds are identical, pitch commonality is defined to take a maximum value of 1. The following formulation of pitch commonality satisfies these two criteria:

$$C = \frac{\sum_P \sqrt{S_1(P) S_2(P)}}{\sqrt{\sum_P S_1(P) \sum_P S_2(P)}} \; . \tag{4.29}$$

Here, S_1 is the array of tone saliences in the first sound, S_2 in the second; and pitch category P is varied over the range of hearing. In the case of two identical sounds, both the numerator and the denominator of the equation are equal to the sum of the tone saliences, i.e. the multiplicity, of each sound.

4.6.2 Pitch Distance

The average apparent *pitch distance* between a pair of sounds is formulated according to the following criteria. (i) The pitch distance between two identical sounds is zero. (ii) The pitch distance between two pure tones (with no subsidiary pitches) is equal to the difference between their frequency levels in semitones. (iii) The pitch distance between different sounds is always greater than zero. A formulation satisfying these criteria is

$$D = \sum_{P} \sum_{P'} S_1(P) S_2(P') \, |P' - P| - \sqrt{\sum_{P} \sum_{P'} S_1(P) S_1(P') |P' - P| \sum_{P} \sum_{P'} S_2(P) S_2(P') |P' - P|} \, , \qquad (4.30)$$

where pitch categories P and P' are varied over the range of hearing.

In a first approximation, the saliences of (simultaneous or sequential) tones falling in different pitch categories may be assumed to be independent. So the product of $S(P)$ and $S(P')$ may be interpreted as the probability that tone sensations at P and P' are both noticed in a particular presentation, i.e. the probability that the interval of size $|P' - P|$ between them is noticed. This applies whether the two tone sensations in question are simultaneous or sequential.

The first term on the right-hand side of (4.30) is a weighted sum of pitch distances between *sequential* tone sensations in a pair of sounds; the second is the geometric mean of weighted sums of pitch distances between *simultaneous* tone sensations in each sound. It is easy to show that the equation satisfies criteria (i) and (ii) above. Verification that it satifies (iii) requires a long mathematical proof [Morris 1986]. Incidentally, criterion (iii) is not satisfied if the second right-hand term is formulated as an arithmetic mean rather than a geometric mean.

Equation (4.30) turns out to be unsuitable for use in music theory. The reason involves the distinction between actual tones (notes) and implied tones (such as Rameau's *basse fondamentale*, Sect. 1.4.1). In music theory, implied tone sensations are invoked to explain harmonic progression (in particular, through the concept of root progression), but they play no role in the theory of voice leading. Voice leading depends on *actual* notes, delineated for the listener by temporal cues such as timbre and onset asynchrony. Good voice leading involves making each melodic line within a chord progression coherent, so that the progression as a whole coheres regardless of which voice dominates (either accidentally or deliberately) in performance.

A more suitable equation than (4.30) for music-theoretical use is obtained by setting $S(P) = 1$ for the actual notes in a pair of chords and $S(P) = 0$ for all the other pitches:

$$D' = \sum_i \sum_j |P_j - P_i| - \sqrt{\sum_i \sum_{i'} |P_{i'} - P_i| \sum_j \sum_{j'} |P_{j'} - P_j|} \,, \qquad (4.31)$$

where i labels the notes in the first chord ($i = 1, \ldots, I$, for a chord with I notes), j labels the notes in the second chord (similarly), and P_i and P_j are the pitch categories of notes in the first and second chords, respectively. This approximation is only valid for small numbers of simultaneous notes: it is difficult to hear more than three tones at once (Sect. 5.2.3), so tones in chords with more than three voices have saliences considerably less than one.

4.6.3 Pitch Analysis Experiment

The results of the pitch analysis experiment were modelled (Sect. 5.3.4) by a combination of pitch commonality and pitch distance, in the special case where one of the two sounds is a pure tone (the probe). Experimental evidence (Sect. 2.4.5) suggests that pure tones have weak subharmonic pitches. Inclusion of these in the model [according to (4.19)] greatly increased calculation time but made very little difference to the fit between results and calculations, so each pure probe tone was assumed to evoke a single tone sensation (with salience $S = 1$) in the pitch category corresponding to its frequency level. Consequently, the pitch commonality (C) of target sound and a probe tone was assumed to be simply

$$C = \sqrt{S(P)} \,, \qquad (4.32)$$

i.e. the square root of the tone salience in the target sound at the pitch category P as the probe tone. The square root in (4.32) comes from the more general formulation of C in (4.29).

The average apparent pitch distance (D) between target and probe in the experiment was assumed to be

$$D = \sum_{P'} S(P') \, |P' - P| \,, \qquad (4.33)$$

where P is the pitch category of the probe tone, P' is the pitch category of any tone sensation in the target sound, $S(P')$ is its calculated salience, and $|P' - P|$ – the absolute value of the difference between P' and P – is the interval between the two pitches in semitones. In the experiment P' varied from 9 to 96 (the range of the modern piano).

The theoretical or calculated outcome of each trial in the experiment was given in relative (unscaled) form by

$$R_t = k_R \frac{C}{\sigma_C} - (1-k_R) \frac{D}{\sigma_D} , \tag{4.34}$$

where k_R – the "relationship perception parameter" – is the fourth and final free parameter in the model, and σ_C and σ_D denote standard deviations over calculated values of C and D for all trials. The parameter k_R reflected the extent to which pitch commonality C, as opposed to pitch distance D, determined the theoretical response R_t. Its ideal value, for those listeners who were able to answer only the question asked of them in the pitch analysis experiment and ignore pitch distance altogether, was $k_R = 1$. Few participants in the experiment were able to listen this analytically, so the optimal value of the parameter was normally considerably less than one.

The theoretical (i.e. calculated) results R_t were scaled against the mean experimental responses (R_e) to each trial as follows:

$$R_t' = \frac{\sigma_e}{\sigma_t} (R_t - \bar{R}_t) + \bar{R}_e , \tag{4.35}$$

where $\bar{R}$ denotes the mean result, and σ denotes standard deviation of results about this mean, calculated over all 158 trials of the experiment. This equation performs a linear transformation on the theoretical responses R_t, setting their mean and standard deviation equal to the mean and standard deviation (respectively) of the mean experimental responses R_e.

4.6.4 Similarity Experiments

In the experiments described in Sects. 5.6 and 5.7, listeners judged the similarity of pairs of tones and chords. The results were modelled by means of the general formulations of pitch commonality and pitch distance in (4.29 and 4.30) above.

In addition, the pitch proximity of two sounds was defined to have a maximum value of 1 for identical sounds, and to approach 0 for sounds which are far apart in pitch relative to pitch distances typical of the prevailing context:

$$X = \exp(-D/\bar{D}) . \tag{4.36}$$

Here, $\bar{D}$ is the mean pitch distance between pairs of sounds, calculated over all trials of the experiment. Note that the effect of context was randomized in these experiments by presenting trials to each listener in a different (random) order.

The theoretical results R_t of the above two experiments were calculated by linear combination of pitch commonality and pitch proximity:

$$R_t = k_R \frac{C}{\sigma_C} + (1-k_R) \frac{X}{\sigma_X} \tag{4.37}$$

where k_R (the relationship perception parameter) is the fourth free parameter in the model [see also (4.34)], and σ denotes standard deviation. Final calculated responses R_t' were scaled against actual mean responses R_e in the same way as for the pitch analysis experiment (4.35).

5. Experiments

The experiments described in this chapter aimed to provide appropriate data by which the model (Chap. 4) could be developed, tested and fine-tuned. In the *multiplicity* experiment, listeners estimated the number of tones in musical sounds (full complex tones, octave-spaced tones, bell-like sounds and chords). The *pitch analysis* experiment measured the saliences of particular tones in musical sounds by a probe tone technique. Four further experiments investigated the apparent *similarity* of sequential pairs of tones and chords of similar loudness, and its dependence on timbral similarity, pitch distance and pitch commonality. Experimental results were used to establish optimal values for three of the four free parameters of the model. These values were subsequently used in music-theoretical applications of the model (Chap. 6).

5.1 General Method

5.1.1 Results and Modelling

Each experiment consisted of a fixed number of trials, and each experimental run (by a particular listener) produced one piece of experimental data for each trial. No results were rejected or eliminated.

The mean response to each trial in an experiment was initially calculated over all runs of the experiment, and 95% confidence intervals were calculated for each mean response. Confidence intervals for different trials normally covered about the same range, so the mean results of trials in an experiment were "significantly different" according to the standard *t test* ($p<0.05$) if they differed by more than the half-width of a 95% confidence interval divided by $\sqrt{2}$.

Results were also calculated for particular groups of listeners in each experiment. The responses of different groups to a particular trial usually had similar deviations, so the size of 95% confidence intervals for individual groups was usually roughly proportional to the reciprocal of the square root of the number of listeners in the group. For example, the confidence interval for a group of 10 was about twice as wide as that for a group of 40. This rule of thumb allowed the significance of differences between results of different groups to be estimated from graphed results (Figs. 5.2–6, 5.8).

Out of six experiments, two (*similarity of piano tones*, *similarity of synthetic tones I*) primarily aimed to produce *qualitative* data. These results inspired the extension of Terhardt's model [Terhardt et al. 1982b] to include formulations of pitch commonality and pitch distance (Sect. 4.6). The results of the other four experiments (*multiplicity*, *pitch analysis*, *similarity of synthetic tones II*, *similarity of chords*) were compared with calculated results according to the model. This allowed the model to be tested and improved.

The correlation between calculations and results was optimized by independent adjustment of the model's four free parameters. Calculations were fitted first to mean results of all listeners, and then to mean results of each group. Each parameter was supposed to reflect how analytically listeners responded to their experimental tasks at a particular level (Sect. 4.1.2). Due to the arbitrariness of the mathematical formulations in the model, and its failure to account for important cultural effects, optimal parameter values should be interpreted with considerable caution. This is particularly true in the case of individual groups, as the mean results of individual groups are based on smaller data samples.

The calculated result for a trial was said to deviate significantly from the experimental result if it lay outside the 95% confidence interval of the mean result for that trial. Such deviations were explicable in terms of the following. Either (i) the experimental result had a cultural or arbitrary component; (ii) there was something wrong with the experimental design; or (iii) psychoacoustical theory had been wrongly applied or inappropriately interpreted. (The same applies for disagreement between calculations and music-theoretical conventions in Chap. 6.) Perhaps the most difficult challenge of this study was to design the experiments and the model in such a way that the probability of options (ii) and (iii) would be minimized.

For reasons of space, not all experimental results are presented here. For further details (including raw data) see Parncutt [1987a].

5.1.2 Cultural Effects

The main effect of conditioning on the results of the experiments was assumed to be familiarity with the harmonic pitch pattern described by the audible harmonics of speech vowels. Those experimental results which could be explained in terms of this auditory universal were described as "sensory". It was not generally possible to distinguish between experimental effects which were "directly sensory" and those which were "indirectly sensory", i.e. conditioned by exposure to aspects of a musical style which in turn had developed under the influence of specific sensory constraints (Sect. 3.1.1). Both "directly" and "indirectly" sensory effects may be described as "sensory in origin".

All experimental participants had had considerable exposure to Western music. The experiments were designed to minimize the effect of such cultural conditioning, in particular through the recognition of musical intervals and chords, so as to isolate sensory effects as far as possible. Techniques to reduce cultural effects involved (i) avoidance of familiar musical contexts, (ii) use of simple experimental tasks not involving musical terminology, (iii) brief presentation of a wide range of sounds in each experiment, and (iv) asking listeners to respond spontaneously. Regarding (ii), musical terminology was avoided in instructions to listeners for all experiments, and musically trained listeners were discouraged from interpreting instructions in musical terms. Regarding (iv), large numbers of trials were needed to reduce the uncertainty and variability ("noise") associated with spontaneous responses.

Effects of individual differences in musical experience were minimized by averaging results across listeners with a range of musical backgrounds. The *tone similarity* experiments (Sects. 5.4–6) were performed in 1983 at the Institute of Electroacoustics, Technical University of Munich, by Western adults with and without musical training. Musicians were mainly students at the Munich Musikhochschule; non-musicians, students and staff at the Institute of Electroacoustics. The *multiplicity*, *pitch analysis* and *chord similarity* experiments (Sects. 5.2, 3, 7) were performed in 1985 at the Department of Psychology, University of New England, Australia. Participants included not only Western adults with and without musical training, but also musically trained Eastern adults (members of a local *gamelan* orchestra) and musically trained Western children (from the local community). With the exception of the children, most participants were students or staff of the University of New England. In all experiments, listeners were designated "musicians" if they played a musical instrument regularly at the time of the experiment.

5.2 Multiplicity

5.2.1 Introduction

Complex sounds presented out of context generally have ambiguous or multiple interpretations (Sects. 2.3.4, 2.4.4). The present experiment was concerned with apparent *multiplicity*, i.e. the number of tones simultaneously noticed, in short, isolated presentations of musical sounds. Apparent multiplicity differs significantly from the actual number of tones, in a similar way that the apparent number of tones in a melody differs from the actual number [Kowal 1987].

Thurlow and Rawling [1959] investigated the accuracy with which listeners could estimate the number of tones in simultaneities comprising one, two or three pure tone components. Responses were found to correlate badly with actual numbers of components. This result is not surprising, as complex tones in speech and music are normally perceived as single entities, even though they contain several audible components (harmonics).

According to Stumpf's [1898] theory of *fusion*, "tones tend to fuse, to interpenetrate each other"; and "degree of fusion is a function of the vibration ratio of the components" [Boring 1942, pp. 360, 361]. In other words, fusion depends on the presence of strong harmonic intervals such as the octave, fifth, fourth and thirds. The above quotations were copied from a publication by DeWitt and Crowder [1987], concerning experiments on the apparent multiplicity of simultaneities of one, two or three *complex* (i.e. musical) tones. Stumpf's theory was confirmed by an analysis of reaction rimes (latencies).

The present experiment aimed to obtain data on the number of tones simultaneously noticed in a wider range of musical sounds. These data were then used to check the validity of eq. (4.12–24), and to establish a rough, general value for the free parameter k_S in (4.24).

Thurlow and Rawling [1959] found that estimates of multiplicity depend on the number of allowed response categories: the more tones listeners are "invited" to hear, the more tones they report. In the present experiment, four response categories were used. This number was expected to be high enough to reflect multiplicities of different sounds relative to each other with reasonable precision, but not so high as to encourage an unusually analytical listening attitude.

5.2.2 Method

Listeners. There were 39 participants in the experiment. Ten were adults with four or more years of training in Western classical music; they were designated "Western musicians". Eight were "Eastern musicians": Indonesian *gamelan* players, who played no Western musical instruments. Nine were "Western non-musicians": adults unable to play any musical instrument, but well acquainted with the sounds of Western popular music. The remaining 12 listeners were "Western children" aged between 10 and 14 years, most of whom were receiving or had received private tuition in classical Western music.

Equipment. Waveform samples were calculated using a Digital PDP 11–23 computer and stored on disc as 12-bit integers. Each sound contained 2000 samples. A 12-bit digital-to-analog (D/A) converter realized the sounds at a rate of 10000 samples per second.

The signal was low-pass filtered using a General Radio Universal filter, type 1952. The filter had a 3-dB cutoff frequency of 2.5 kHz (D_7) and a spectral gradient of −30 dB per octave. A single Pioneer CS-T3 loudspeaker (with tweeter disconnected) played the sounds inside a 4 m × 3 m × 3 m semi-anechoic chamber. The loudspeaker was placed close to one wall at the height of the (seated) listener's head. The frequency response of the filter, amplifier, speaker and room was accounted for when modelling the results by adding the response characteristic of the filter/amplifier/speaker/room system (measured over the frequency range of the sounds using a Brüel & Kjaer sound level meter) to the spectra by which waveshapes had been calculated.

Sounds. The ten sounds investigated both here and in the next (pitch analysis) experiment are notated musically in Fig. 5.1. The frequencies of all pure tone components of all sounds were tuned to the standard equally tempered scale. No component lay outside the range 50–4200 Hz (G_1 to C_8).

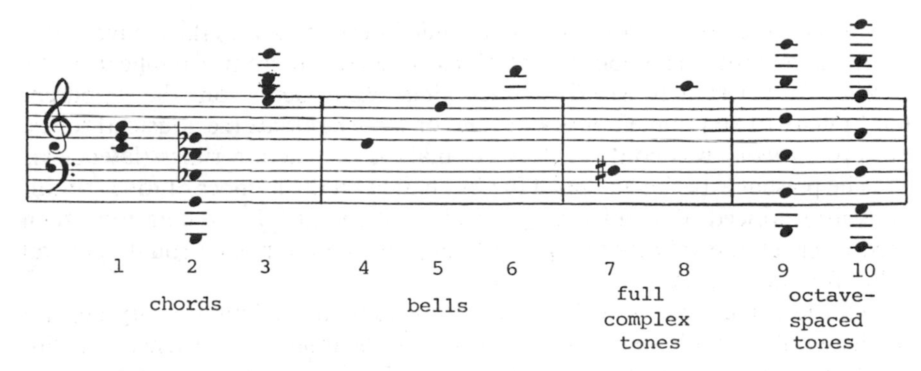

Fig. 5.1. Music notation of sounds presented in the multiplicity and pitch analysis experiments.

The first three sounds were musical chords. Sound number 1 was a C major triad in root position in the middle pitch register, sound 2 was an E flat major triad in first inversion in the bass region, and sound 3 was an E minor triad in root position in the treble. Each note was realized as a complex tone with all harmonics up to the tenth (or the upper cutoff at C_8), and a spectral envelope gradient (before amplification) of -10 dB per octave. The overall sound pressure levels (SPLs) of the complex tone components had been adjusted so that their separate loudnesses (in isolation) were approximately equal. SPLs of coinciding components were combined by adding intensities (4.11).

The next three sounds were bell-like in spectral content. (They didn't sound very much like bells, however, due to their uncharacteristically smooth, noiseless envelopes.) They each had 8 pure tone components at intervals of 0, 12, 15, 19, 24, 28, 31 and 36 semitones above the pitches D_4, D_5 and D_6 (respectively), imitating the superposed major and minor triads in the spectrum of an "ideal" bell [Lloyd 1954; Lehr 1986]. Component SPLs were equal (before amplification).

Sounds 7 and 8 were single harmonic complex tones with musical note names $F\#_3$ and A_5, with spectral envelope gradients (before amplification) of -10 dB per octave. Sound 7 contained all harmonics up to the tenth; sound 8 had only four harmonics below the upper cutoff at C_8. Sounds 9 and 10 were octave-spaced tones on B and F, with flat spectral envelopes (before amplification).

Sounds had equal maximum pressure amplitudes before filtering and amplification. Overall SPLs of the sounds (in order from 1 to 10) were 64, 61, 64, 65, 64, 62, 62, 70, 64 and 64 dB. Differences in loudness were small: for example, sound 8 had the highest SPL (70 dB), but was not noticeably louder than the other sounds, as it comprised only four pure tone components and so covered relatively few critical bands (Sect. 4.3.1).

Sounds had total durations of 0.2 s. The first and last 0.02 s of the sounds were shaped to prevent onset and offset clicks.

Procedure. In each trial, one of the ten sounds (duration: 0.2 s) was presented twice, with a pause of 0.6 s between presentations. After this, listeners could take as long as they wished to respond. They were asked, however, to respond quickly and spontaneously. The task was to answer the question: "*How many tones do you hear in the sound?*", and possible responses were 1, 2, 3 or 4. No feedback was given.

The sound presented in each trial was shifted up or down as a whole through a randomly chosen distance in the continuous range -2 to $+2$ semitones. This was done by adjusting the rate at which sample values were realized by the D/A converter. The size and direction of the shift was determined independently for every trial, and shifts were different in every experimental run. They were intended to confuse any listeners with latent or unrecognized perfect pitch [see Terhardt and Seewann 1983].

The experiment contained only 20 trials: two presentations of each of the ten sounds, so the probability that listeners became familiar with (and subsequently recognized) any of the sounds was low. The trials were presented in a random order which was different for each experimental run (i.e. for each listener).

5.2.3 Results

Results are summarized in the upper panel of Fig. 5.2. Sound 2, the low five-note chord, had the highest apparent multiplicity (between 3 and 4). Sound 8, the high single tone, had the lowest (between 1 and 2). The three chords (sounds 1 to 3) had significantly different multiplicities from each other. Among the bell-like sounds (sounds 4 to 6) and the full complex tones (sounds 7 and 8), there was a tendency for high-pitched sounds (sounds 6 and 8) to have lower multiplicity than similar sounds in middle or low ranges. Results for the octave-spaced tones (sounds 9 and 10) were not significantly different from those of sound 3 (the high chord) and sounds 4 and 5 (the low and mid-range bells).

The mean responses for each group of listeners are plotted in the lower panel of the figure. Western musicians (triangles) covered the widest range in their responses. This is typical in subjective ratings of musical sounds [cf. Roberts 1986], and reflects musicians' greater confidence in their responses. Western musicians also gave the highest overall mean result, reflecting their relatively analytical listening attitude.

5.2.4 Modelling

Calculations according to (4.24) were fitted directly to the mean responses for each group. Values of the free parameters were adjusted for optimal fit between calculations and results (Sect. 5.1.1). The parameters k_M and k_S in (4.14, 24) took fairly typical values: 29 and 0.51 respectively. The tone perception parameter k_T in (4.19) was quite high (5.6 in comparison to the more usual

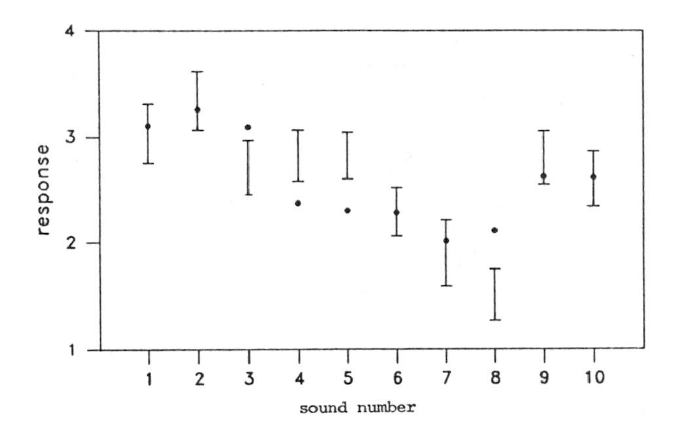

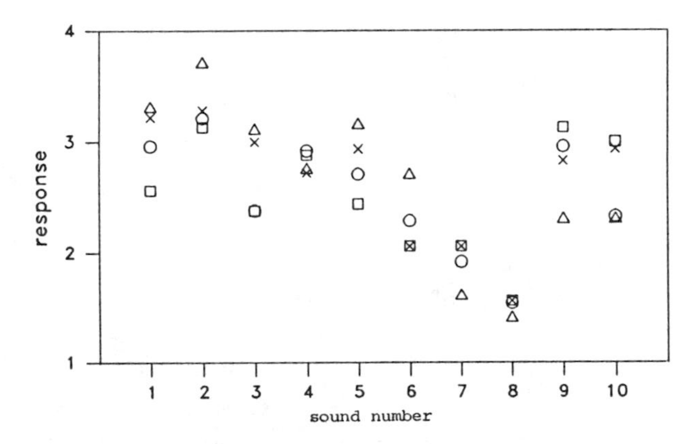

Fig. 5.2. Results of the sensory multiplicity experiment. *Upper panel*: 95% confidence intervals of mean responses of all 39 listeners, 2 data each (*bars*); calculated results according to (4.24) with $k_M = 29$, $k_T = 5.6$ and $k_S = 0.51$ (*points*). *Lower panel*: mean results of 10 Western musicians (*triangles*), 8 Eastern musicians (*squares*), 9 Western non-musicians (*crosses*) and 12 Western children (*circles*)

value of 3; see Sect. 5.3.4), suggesting that pure tone components were "heard out" during the experiment, perhaps because of the analytical bias of the experimental task.

Free parameters were also adjusted to fit the results of individual groups (Sect. 5.1.1). The masking parameter k_M was set higher (to a value of 50) for Western musicians than for the other listeners, possibly reflecting an ability to infer the pitches of inaudible (i.e. masked) components of a familiar musical sound. The simultaneity perception parameter k_S was set to slightly higher values for Western musicians (0.52) and non-musicians (0.53) than for the other listeners, in order to fit their slightly higher responses.

Calculations and results are compared for all listeners in the upper panel of Fig. 5.2. An initial observation is that the range of the calculations is smaller than that of the responses. This may be due either to the imprecision of the model, or to the tendency of listeners to adapt their responses to fill the allowable range (in this case, 1–4). Calculated multiplicities tended to be too high for high-pitched sounds (sound 3, 6 and 8), apparently because of the failure of (4.17, 19) to account for spectral dominance (Sect. 6.1.2). Responses tended to be higher than calculations for the bell-like sounds (4, 5 and 6): listeners may have concluded from their relatively strange sound that relatively many components had been needed to synthesize them.

5.2.5 Conclusions

In agreement with Thurlow and Rawling [1959], the mean apparent number of simultaneous tones (multiplicity) of musical sounds was found to differ from the number of pure tone components. It also differed from the number of (full, harmonic) complex tone components. For example, the result for a single complex tone in register 3 (sound 7) was surprisingly high (almost 2), both experimentally and according to the model.

The formulation for multiplicity used in the model failed to account for some experimental effects (e.g. spectral dominance). The model nevertheless seems sufficiently logical and accurate for more qualitative music-theoretical applications.

The simultaneity perception parameter k_S in (4.24) was estimated at 0.5. This value is used in music-theoretical applications of the model (Chap. 6).

5.3 Pitch Analysis

5.3.1 Introduction

The previous experiment was concerned with how many tones an average or typical listener notices simultaneously in a musical sound. Due to the pitch ambiguity of musical sounds (Sect. 2.4.4), different listeners can notice different tones or sets of tones in the same sound, and the same listener can notice different sets of tones when the sound appears at different times and in different contexts. Clearly, it is impossible to predict exactly which tones a particular listener will notice in a particular situation. The best one can do is predict the *probability* that a given tone in a given sound will be perceived. The present experiment aimed to estimate such probabilities (or *saliences*) as a function of pitch, for the ten sounds of the previous experiment.

Terhardt et al. [1986] investigated the pitch ambiguity of complex tones by a pitch-matching procedure: listeners were asked to adjust the frequency of a pure tone until its pitch coincided with that of a complex sound. This method has the advantage that listeners are free to choose probe pitches, so responses are not biased in favour of particular pitches.

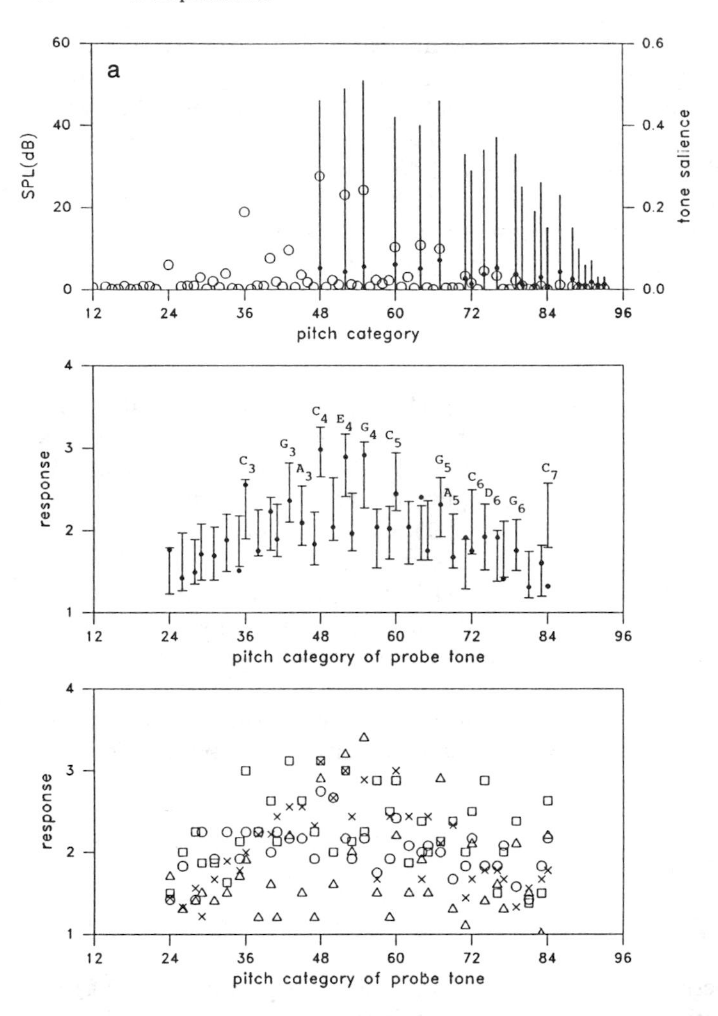

Fig. 5.3a, b, c, d. Results of the pitch analysis experiment. *Upper panels*: Pure tone components (*vertical lines*, *left scale*), pure tone sensations (*points, right scale*) and complex tone sensations (*circles, right scale*). *Centre panels*: 95% confidence intervals of mean responses of all 39 listeners (*bars*); calculations according to (4.35) with $k_M = 45$, $k_T = 3.5$, $k_S = 0.4$, and $k_R = 0.65$. *Lower panels*: Mean results of 10 Western musicians (*triangles*), 8 Eastern musicians (*squares*), 9 Western non-musicians (*crosses*) and 12 Western children (*circles*). (**a**) Sound 1: Chord $C_4 - E_4 - G_4$

The present experiment used a slightly different method. The frequencies of probe tones were set in advance, and listeners were asked whether or not each probe sounded like it was part of the sound which preceded it. This allowed salience values to be compared across different sounds.

The *pitch weights* of a complex sound according to Terhardt's model [Terhardt et al. 1982b] are relative estimates of salience: they may only be compared within sounds, not across sounds. For example, the pitch salience of a

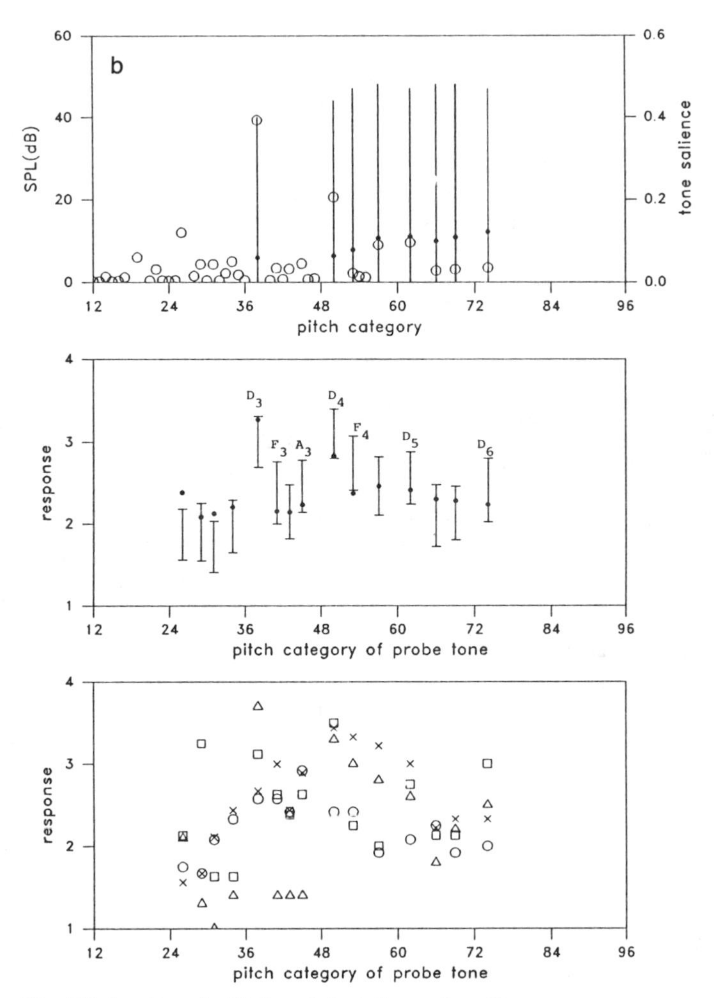

Fig. 5.3 *(continued).* (**b**) Sound 4: Bell sound on D_3

pure tone is slightly greater than that of a complex tone [Fastl and Stoll 1979]. According to Terhardt's model, however, the calculated pitch weight of a clearly audible pure tone is 0.5 (when spectral pitch weights are reduced to 50% so they become comparable with virtual pitch weights), while the calculated weight of the main pitch of a full complex tone in the pitch range of speech lies in the range 2.5–3. The present experiment's method of setting probe frequencies in advance enabled the development and testing of a model by which saliences are estimated as absolute values, see (4.25).

5.3.2 Method

Listeners and **equipment** were the same as in the multiplicity experiment (Sect. 5.2.2).

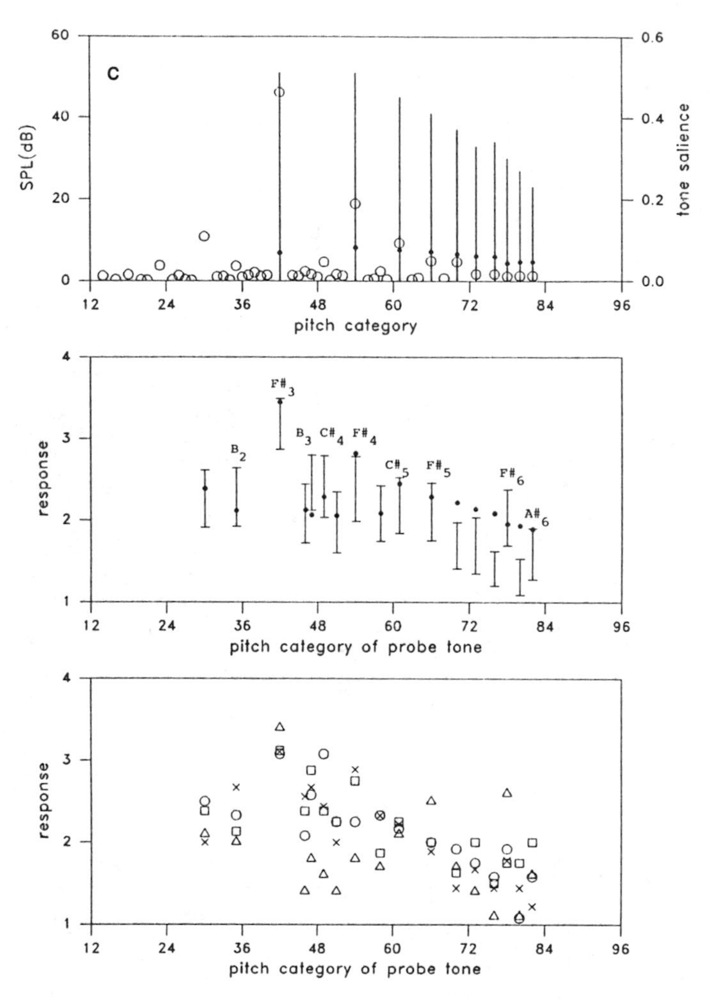

Fig. 5.3 *(continued).* **(c)** Sound 7: Full complex tone on F$\#_3$

Sounds. The same ten complex sounds were investigated as in the previous experiment. The sounds had been composed such that their complex tone components were fairly evenly distributed across the central pitch range and around the chroma cycle; otherwise, undue emphasis on particular pitches or chroma during the experiment could have influenced the results via serial (context) effects.

In addition, pure *probe tones* were presented. Probe tones had frequencies in the range 65 Hz–2.1 kHz (C_2-C_7). They corresponded to tone sensations that had been predicted by an early version of the model to be evoked with saliences greater than some arbitrary minimum value. Sound 1 was additionally compared with probe tones whose pitches were not, according to the model,

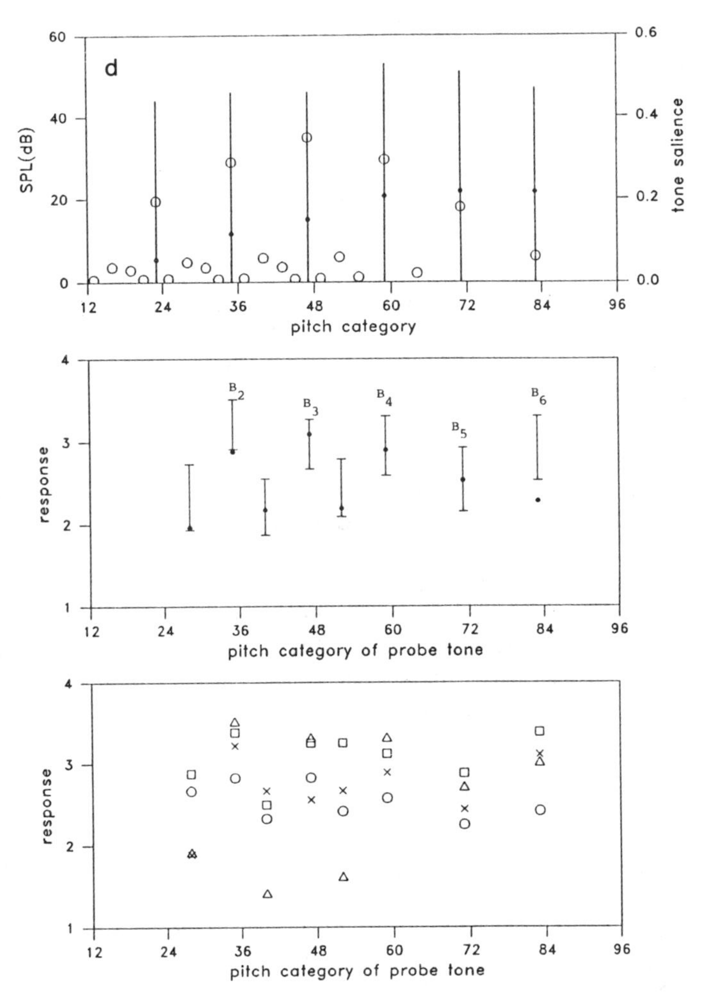

Fig. 5.3 *(continued).* **(d)** Sound 9: Octave-spaced tone on B

evoked at all (Fig. 5.3a): in all, it was compared with 36 different probe tones, representing five octaves of the C major scale. The numbers and pitches of probe tones presented in conjunction with sounds 4, 7 and 9 can be seen from Figs. 5.3b–d.

The SPLs of probe tones were adjusted for approximately equal loudness. Probe tones were consistently quieter than the sounds being analyzed, but still clearly audible, in order to make it easier for listeners to imagine a probe tone to be "part of" a sound.

The frequency of each probe tone was tuned by the author, by comparing it with the complex sound with which it was to be presented. The adjustment process was performed in 0.1 semitone steps. Some probe tones had to be adjusted away from their original "equally tempered" frequencies by a semitone or even more. Pitch shifts of this magnitude are not unusual [Walliser 1969c]. In retrospect, it would have been preferable to ask a number of different musicians to tune the probe tones and average the results, as pitch shifts vary from one person to another.

Procedure. Each trial began with a complex sound (0.2 s), followed by a pause (0.3 s), then a probe tone (0.2 s), a longer pause (0.6 s), and the same sound-pause-probe group repeated (0.2 + 0.3 + 0.2 s). After this, listeners could take as long as they wished deciding on their responses, but were encouraged to do so quickly and spontaneously. Each sound-probe pair was shifted through a random interval in the range −2 to +2 semitones, in order to minimize the chance of perfect pitch effects influencing the results.

Listeners were asked "*Does the tone* (i.e. the second sound) *sound like it is part of the first sound?*" They responded by pressing one of four buttons, labelled (from left to right): "No", "No (not sure)", "Yes (not sure)", and "Yes". These four responses were mapped onto the numerical values 1, 2, 3 and 4 (respectively) for the purpose of recording responses and graphing data. Listeners were free to use as many or as few buttons as they wished. It was stressed that responses would be interpreted as opinions – that there were no absolutely right or wrong answers with which responses could be compared.

The experiment contained 158 different trials, presented in a unique random order for each listener. Listeners took a short break after the 80th trial.

5.3.3 Results

Overall results are displayed in the centre panels of Figs. 5.3 for sound 1 (C major triad), sound 4 (bell on D_3), sound 7 (full complex tone of $F\#_3$) and sound 9 (octave-spaced tone on B). In general, responses tended to be highest at pitches corresponding to actual (pure or complex) tone components, and to fall off with increasing pitch distance from actual tone components. This suggests that the mean response for a particular trial depended not only on the salience of the tone sensation at the pitch of the probe but also on the proximity of other tone sensations, even though pitch proximity was not implied by the question asked of the listeners. The musical note names in the centre panels of the figures indicate tone sensations in the sounds whose psychological reality was demonstrated ($p < 0.05$) by comparing results for neighbouring probe tones (Sect. 5.1.1).

The mean responses for each group of listeners are plotted in the lower panels of Fig. 5.3. As in the multiplicity experiment, the responses of the Western musicians covered the greatest range, reflecting their confidence. The Western and Eastern musicians often disagreed with the other listeners (and with each other), presumably due to their musical training (i.e. cultural influences). Responses to trials testing the existence of "residue tone sensations" (complex tone sensations not corresponding to actual pure tone components) were generally lowest for the Western musicians. This was to be expected, as the musical sounds with which Western musicians are familiar usually comprise notes with audible fundamentals.

Eastern musicians (squares) gave significantly higher responses than the other groups (Sect. 5.1.1) for many of the tested residue tone sensations (e.g. C_3 and G_3 in sound 1, F_2 in sound 4, and E_4 in sound 9). This may have been due to their experience with non-harmonic sounds such as *gamelan* bells. The spectrum of a bell-like sound (Sect. 5.2.2) typically contains several components which conform fairly closely to a harmonic series of frequencies, as well as a few components which do not. For example, most of the pure tone components of the C major triad (sound 1) conformed approximately to a harmonic series on C_3; only the components at E_4 and B_5 did not. So it is possible that sound 1 seemed more bell-like than chord-like to the Eastern musicians.

The three musical chords (sounds 1–3) were found to evoke tone sensations both higher and lower in pitch than the actual notes of the chords. Lower-pitched tone sensations were always of the complex variety, and corresponded to possible roots of the chords in music theory. Higher-pitched tone sensations were either pure (heard-out harmonics) or complex.

The three bell-like sounds (sounds 4–6) and the two full complex tones (sounds 7 and 8) each evoked more than one tone sensation, as expected from the results of the multiplicity experiment. If only one tone had been noticed in these sounds, then its pitch would have been ambiguous. The two most salient tone sensations were generally octave equivalent.

The octave-spaced tones (sounds 9 and 10) were found to evoke tone sensations at all pitches corresponding to actual tone components. Due to the limited selection of probe tone pitches, the prediction (Sect. 6.1.5) that subfifth tone sensations (E in sound 9, B*b* in sound 10) are also evoked was not directly tested. However, the result for the Eastern musicians at E_4 in sound 9 suggests that at least some of the listeners noticed subfifths in octave-spaced tones.

5.3.4 Modelling

Results were modelled according to (4.12–35). Equation (4.35) transformed the set of 158 calculated responses in such a way that their overall mean and standard deviation were set equal to the overall mean and standard deviation of the mean experimental responses. Optimal values of the four free parameters in the model were as follows. The masking parameter k_M in (4.14) was relatively high at 45 (instead of a more usual 25), suggesting either that listeners detected pure tone components that, according to Terhardt [1979a], are inaudible, or that relatively high values were needed in order to compensate for the inaccuracy of the model's masking algorithm, in comparison to Terhardt's. The value of k_T in (4.19) was 3.5: this was probably slightly higher than normal, due to the analytical bias of the experimental task. The value of k_S in (4.24) was a little low (0.3 as opposed to 0.5). The value of k_R in (4.34) was relatively high (0.65), reflecting the emphasis on pitch commonality in the experimental task.

Parameter values were also adjusted to fit the results of each listener group. The tone perception parameter k_T was highest (6) for the Western musicians (implying that they had the greatest tendency to hear out pure tone components), and lowest (2.5) for the Western children. The simultaneity perception parameter k_S was also highest (1.1) for the Western musicians (reflecting their ability to notice several tone sensations at once in a musical chord), and lowest (0.1) for the Western children. The pitch relationship parameter k_R was highest (0.8) for the Western musicians (i.e. they were most able to ignore pitch distance and concentrate on pitch commonality); it was lowest (0.4) for the Western non-musicians.

Calculations failed to fall within 95% confidence intervals of overall mean responses in 31 of 158 trials. If the model had adequately accounted for all relevant effects, then only about 8 calculated responses (5% of 158) would have been expected to fall outside the 95% confidence intervals. Clearly, the model did not account for all significant experimental effects.

Possible reasons why mean responses to certain trials were significantly *lower* than calculated responses are: that listeners knew from experience (especially of music) that a probe tone was not "really" in a sound, even though it "sounded like" it could be; and that the probe tone sounded out of tune with the corresponding tone sensation evoked by the sound. The above effects occurred most regularly for the Western musicians, as expected from their experience and training. Possible reasons why some mean responses were significantly *higher* than calculated responses are: that the pure tone component at the pitch of the probe tone was "heard out"; that the probe tone was heard to belong to the same category as a salient tone sensation one semitone higher or lower; that the probe tone was octave equivalent to a salient tone sensation; and that the probe tone corresponded to the lowest or most prominent pure tone component normally heard in a familiar, non-harmonic sound (e.g. a bell), but which was not necessarily present in the actual sound.

5.3.5 Conclusions

In agreement with Terhardt et al. [1986], typical musical sounds were found to evoke tone sensations corresponding neither to musical notes nor to pure tone components. Such tone sensations generally stand in strong harmonic relationships with actual notes and pure tone components. This finding supports Terhardt's theory that harmonic relationships in music theory have a sensory basis.

Formulas for "pitch weight" in Terhardt's model [with minor alterations, (4.17 and 4.19)] adequately reflected the relative relative audibilities of (pure and complex) tone components within complex sounds. A new addition to the model (4.25) enabled the saliences of tone sensations to be expressed as absolute values and compared across different sounds. Overall pitch distances between probe tones and experimental sounds (4.33) also influenced the results of the experiment. Agreement between the calculations of the model and mean experimental responses was satisfactory, considering that many effects that appear to have influenced the results were not accounted for at all by the model.

The results of the modelling procedure suggested a typical value of 3 for the tone perception parameter k_T. This value is used in music-theoretical applications of the model (Chap. 6).

5.4 Similarity of Piano Tones

5.4.1 Introduction

Terhardt [1983] suggested a sensory basis for the harmonic relationship between sequential complex tones at musical intervals such as octaves, fifths and fourths (Sect. 3.2.3). He proposed that such tones were to some extent "confusable" with each other due to the octave- and fifth-ambiguity of their pitch. If this is true then sequential tones at these intervals should sound more *similar* to each other than sequential tones at slightly different intervals.

Thurlow and Erchul [1977], in their "piano octave-similarity tests", asked listeners to rate the similarity of sequential piano tones. The difference between the similarity of sequential tones an octave apart and the similarity of tones spanning minor sevenths and major ninths was only significant for 3 out of 9 listeners, 8 of whom could recognize octaves. This result does not necessarily argue against Terhardt's proposal. According to music theory, tones a minor seventh or major ninth apart are not harmonically remote from each other: they lie only two steps apart on the cycle of fifths. So they, like tones an octave apart, may have some harmonic affinity.

In the present experiment, similarity of piano tones was compared for octaves and *major* sevenths (12 and 11 semitones), corresponding to spans of zero and five steps respectively on the cycle of fifths. The experiment therefore provided a more sensitive test of the octave-similarity of piano tones that of Thurlow and Erchul. In addition, the fifth relationship was tested, by comparing similarity ratings of fifths and tritones.

5.4.2 Method

Listeners. A total of 22 musicians and 12 non-musicians took part, see Sect. 5.1.2.

Equipment. The tones presented in the experiment had been played by the author on a Steinway grand piano ($A_4 = 442$ Hz) and tape recorded. The piano was about three years old and in excellent condition. No pedal was used. Tones were played as far as possible with the same loudness. Timing was controlled by means of an electronic metronome with an earplug. A condenser microphone (Peerless MBC540) and reel-to-reel tape recorder (Revox A77, 19 cm/s) were used to make a monophonic recording of the tones.

The timbre of selected tones on the tape was changed by electronical filtering. One filter (built at the Institute of Electroacoustics, Technical University of Munich) had a lowpass characteristic, cutting off very steeply (roughly 200 dB/octave) at 840 Hz ($A\flat_5$). The other had a bandpass characteristic with 36 dB/octave flanks at 600 Hz (D_5) and 2 kHz (B_6). Levels were adjusted so that filtered and unfiltered tones had roughly the same loudness.

The listener sat in a sound-isolated both, and heard the tones diotically (same in each ear) over electrodynamic headphones (Beyer DT48) whose frequency response was compensated for by a free-field equalizer [Zwicker and Feldtkeller 1967].

Sounds. Tones on the final tape were either "unfiltered", "low-pass filtered", or "high-pass filtered" (actually, band-pass filtered, as described above). The levels of the tones were adjusted so that all were moderately quiet, but still clearly audible (approximately 60 dB SPL).

Sixty-two tone pairs were tested for similarity. Of the 62 tone pairs, 16 consisted of two unfiltered tones encompassing the intervals 1, 4, 6, 7, 11, 12, 16 and 19 semitones, of which either the upper or the lower tone was middle C; 6 consisted of two middle C's of different timbre; and the remaining 40 involved combinations of the above intervals and timbres.

Procedure. The tone pair in each trial was presented twice in succession. Each tone had a duration of about 0.4 s. The time interval between the tones in a pair was about 0.4 s; between pairs, 1.0 s; and between trials, 4.6 s.

During the time interval between trials, the listeners rated the *similarity* of the two tones on a 4-point scale, labelled "sehr unähnlich . . . sehr ähnlich" ("very dissimilar . . . very similar"). They were asked to regard similarity as a general term concerning all properties of the tones. Responses were written on a prepared form.

Each trial was presented twice, in two different orders. Altogether, 124 trials were presented, in a random order that was the same for each listener.

5.4.3 Results

Results for the 16 pairs of unfiltered tones are illustrated in Fig. 5.4. The responses for each pair are averaged over two presentations (in two different orders) to each listener.

Consider first the mean results of all 34 listeners. They indicate clearly that tones spanning an octave (12 semitones) sound more similar to each other than tones spanning a major seventh (11). In addition, tones spanning a perfect fifth (7 semitones) were rated more similar than tones spanning a tritone (6). The latter effect was smaller, but still significant.

For non-musicians, the result for the falling major seventh was not significantly different from the result for the falling octave (Sect. 5.1.1). On the other hand, the results for the rising major seventh and octave *were* significantly different. A possible reason for this involves the effect of context on pitch am-

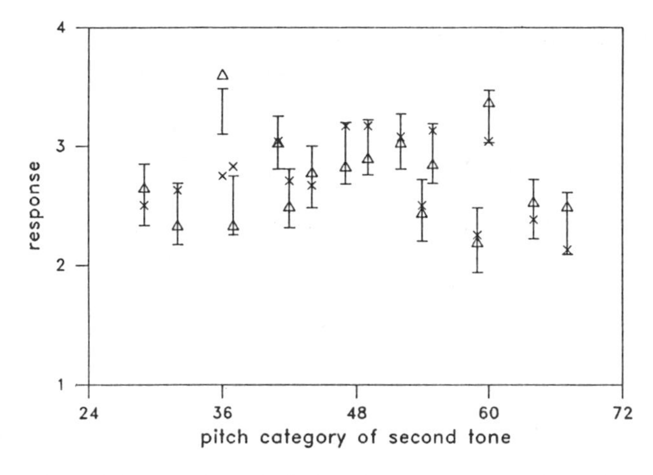

Fig. 5.4. Results of the similarity of piano tones experiment for unfiltered tones. The 95% confidence intervals of mean responses of all 34 listeners, 2 data each (*bars*); mean responses of 22 musicians (*triangles*) and 12 non-musicians (*crosses*). The first tone was always C_4 (48)

biguity. The first of a pair of tones in an experimental trial provides a context for the perception of the second tone. This reduces the pitch ambiguity of the second tone, i.e. it reduces the salience of its subsidiary pitches. The subsidiary pitches of the first tone in a pair should therefore contribute more to the similarity of the pair than those of the second tone. The main subsidiary pitch of a complex tone in the low and middle musical pitch ranges lies an octave above its main pitch [Terhardt et al. 1986] (Sect. 6.1.5). Therefore pitch commonality can be expected to be stronger for rising octave leaps than for falling octaves. This order effect was not accounted for in the model.

The non-musicians apparently did not recognize the octave interval, and so could not distinguish it from the major seventh (although some of them may have succeeded in recognizing octaves of they had been *required* to do so). If they did not recognize the falling octave it is unlikely that they recognized any other interval. It may be concluded that the experimental effects at the rising octave and the rising and falling fifths for non-musicians were predominantly or completely sensory in origin (i.e. due to pitch commonality).

The size of the effect at the fifth was about the same for both groups of listeners. This is consistent with the theory that the responses of the musicians to these trials were determined by the same sensory effects which determined the responses of the non-musicians. The octave effect, by contrast, was bigger for musicians. This difference may be attributed to the cultural component of octave equivalence in music (Sect. 3.3.1). Culturally conditioned octave equivalence apparently only influenced the results of musicians, because – presumably – only they could recognize octaves during the experiment.

The tones of falling intervals were rated significantly more similar than those of rising intervals, for both musicians and non-musicians. This effect may have been due to pitch (pitch distances in the bass are slightly smaller than in the treble, even for complex tones, see Sect. 2.5.2) or timbre (timbre of piano tones does not necessarily vary linearly with pitch). Alternatively, it could have been due to cultural conditioning: descending intervals significantly outnumber ascending intervals in Western popular and art melodies [Jeffries 1974].

Only the results for the 16 pairs of unfiltered tones are presented graphically, as only these are relevant to the aim of the experiment. The other tone pairs were included only to encourage listeners to attend to timbral variations, and thereby to reduce the likelihood that they would recognize intervals and respond according to musical knowledge.

5.4.4 Grouping

The 34 listeners were classified into groups according to the extent to which harmonic relationship, pitch proximity and timbral similarity influenced their responses. (No such objective classification has been possible in the multiplicity and pitch analysis experiments, due to lack of simple grouping criteria.) Listeners were assigned to a particular group if they rated the similarity of certain tone pairs to be, on average, at least one rating scale category higher than the similarity of certain other tone pairs. A *harmonic relationship* group was established by comparing responses for octaves and fifths with responses for major sevenths and tritones, between unfiltered tones; 10 musicians and 1 non-musician were allocated to the group. A *pitch proximity* group was established by comparing responses for twelfths and tenths with responses for fifths and thirds (again between unfiltered tones); the group contained 5 musicians and 5 non-musicians. A *timbral similarity* group was established by comparing the responses for the 16 unfiltered pairs those for the 46 filtered pairs; it contained 11 musicians (i.e. all but one) and 5 non-musicians. Of the three groups, the harmonic relationship group had the highest ratio of musicians to non-musicians, due to the particular emphasis placed on octaves by the musicians. The timbral similarity group was the largest of the groups, suggesting that timbral similarity was the most salient effect in the experiment, and supporting the claim that relatively few listeners recognized musical intervals.

5.4.5 Conclusions

Similarity ratings of pairs of tones of different pitch and timbre were determined mainly by similarity of timbre, and – to a lesser extent – by pitch distance and harmonic relationship. Pairs of musical tones an octave or a fifth apart were usually more similar ($p<0.05$) than pairs covering chromatically neighbouring intervals, even for non-musicians, who apparently did not recognize musical intervals at all during the experiment. This is consistent with the theory that the similarity of such tones is sensory rather than cultural in origin. On the other hand, the large difference in the similarity of tones spanning octave and major seventh intervals as perceived by musically trained listeners appeared to be mainly due to cultural conditioning.

5.5 Similarity of Synthetic Tones I

5.5.1 Introduction

Kallman [1982] compared *pure* tones in musical intervals for similarity. In general he found no effect at the octave. Similarly, Thurlow and Erschul [1977], in their "tone octave-similarity test", found that listeners either recognized octaves between pure tones and rated similarity on that basis, or else they showed no significant octave effects at all. These results contrast with the results of experiments on similarity of *complex* tones [Stoll and Parncutt 1987] (Sect. 5.4), in which significant effects at the octave were observed even in the (apparent) absence of octave recognition.

The present experiment aimed to test whether the above distinction between the similarity of pure and complex tones is general and independent of experimental method. To achieve this aim, pairs of pure tones and pairs of complex tones were presented at different (random) points in a single experimental sequence.

Previous experiments on tone similarity (including Stoll and Parncutt [1987]) had tested a restricted number of musical intervals. In the present experiment, all chromatic intervals from 0 to 13 semitones were investigated. This was intended to reduce the probability of intervals being recognized, and to produce more comprehensive results of similarity as a function of interval.

5.5.2 Method

Listeners. The aim of the experiment was to explore the sensory basis of interval relationships; results in the absence of musical interval recognition were of particular interest. Therefore, more non-musicians than musicians were tested. In all, 9 musicians and 21 non-musicians took part.

Equipment. Waveform samples were generated by a microprocessor system, and converted to analogue signals by a 12 bit D/A converter. They were then low-pass filtered to a 3-dB cutoff frequency of 3.7 kHz (Bb_7) with a filter slope of about 200 dB/octave, using a filter built at the Institute of Electroacoustics, Technical University of Munich. The listener was seated in a sound-isolated booth equipped with a response keyboard, and heard the tones diotically (same in each ear) over headphones via a free-field equalizer.

Sounds. Tones were either pure (with sinusoidal waveforms) or complex (with sawtooth or rectangular waveforms). (The spectra of both sawtooth and rectangular waveforms have spectral envelope gradients of -6 dB per octave; sawtooth waveforms have all harmonics, while rectangular waveforms have only odd-numbered harmonics.) The bandwidth of the complex tones was limited to the lower 16 harmonics, or to the cutoff frequency of the filter (3.7 kHz). Waveforms were sampled at a rate of 18 kHz. The SPL of the pure tones was about 70 dB. Tones were adjusted for roughly equal loudness by setting the SPL of sawtooth tones 12 dB lower, and of rectangular tones 15 dB lower, than that of the pure tones.

Altogether, 60 different tone pairs were tested, of which 3 had the same pitch and the same timbre (i.e. the same waveform), 6 had the same pitch and different timbre, 26 had different pitch and similar timbre (i.e. the same waveform), and 25 had different pitch and different timbre. Fundamental frequencies were tuned to the equally tempered scale with $A_4 = 440$ Hz, and frequencies of harmonics were exact multiples of fundamental frequencies. Of the 26 trials with tones of different pitch and similar timbre, 13 incorporated pairs of pure tones (over intervals $1-13$ semitones) and 13 incorporated pairs of sawtooth tones (over the same intervals). Fundamental frequencies ranged from 131 Hz (C_3) to 523 Hz (C_5).

In the 25 trials with different pitch and timbre, one tone was always a square wave, the other was either sinusoidal or sawtooth, and the interval between the tones was chosen at random from the range $1-13$ semitones. As before (Sect. 5.4.2), these trials were included only to increase emphasis on timbre and thereby reduce the likelihood that intervals would be recognized.

Procedure. Each trial consisted of a repeated tone pair, i.e. of four sequential tones. Each tone had a duration of 0.2 s. The time interval between tones in a pair was 0.15 s; between pairs in a trial, 0.35 s. Listeners had as long as they wished to respond, but they were asked to do so quickly and spontaneously. They rated the similarity of the tones by pressing one of six buttons, labelled "very similar ... very dissimilar". Responses were recorded as digits in the corresponding range $1-6$.

Each experimental run contained 120 trials, in which the 60 different tone pairs were presented in two different orders. The overall order of the experiment was random, and differed for each experimental run. Each listener performed the experiment twice. In one run, pairs of tones were anchored to a particular pitch. In anchored runs, the pitch of one of the tones in a pair was always C_4 (for pure tones) or $F\#_3$ (for sawtooth tones). In the other run performed by each listener, the pitches of both the tones in a pair changed from one trial to the next. In the 9 trials with tones of the same pitch, that same pitch varied from $F\#_3$ to $F\#_4$, and in the 51 trials comprising tones of different pitch, the lower tone varied from C_3 to B_3. Half the listeners performed the anchored experiment first; the others started with the unanchored experiment.

5.5.3 Results

Results for pairs of tones with similar timbre are presented in Fig. 5.5. The plots compare results for pairs of pure (sinusoidal) and complex (sawtooth) tones. Results for pairs of different timbre are not graphed, as these were not relevant to the aims of the experiment; these pairs were only included to encourage listeners to concentrate on timbral similarity, and thereby to reduce the probability that they would recognize musical intervals.

The results for tone pairs which were identical in both pitch and timbre are denoted in the figures by "interval = 0". Almost all listeners responded "very similar" (6) to these trials. This may be regarded as a trivial case of categorical perception (the tones clearly fall in the same pitch category) and of pitch commonality [the tones, according to (4.29), have a pitch commonality of 1].

A feature of all panels of Figs. 5.5 is the effect of pitch distance (interval size). Similarity tends to fall with increasing pitch distance. Responses fall particularly suddenly as pitch distance increases from 0 to 1 semitone, i.e. as pitch proceeds from "same" to "different". The position of this category boundary is determined more precisely in the next experiment (Fig. 5.6a, Sect. 5.6.3).

In Fig. 5.5a, responses for 9 musicians are compared with those for 21 non-musicians. No significant effect occurs for non-musicians and pure tones at any interval except unison, in agreement with Kallman [1982]. This suggests that the effect of pitch commonality on similarity of pure tones (due to weak subharmonic pitches, see Sect. 2.4.5) is negligible. For musicians, on the other hand, there was a strong effect at the octave: they presumably recognized octaves, and rated them according to their "equivalence".

The responses for complex tones have more structure than those for pure tones. Consider first the responses of the non-musicians. The mean responses for intervals of 1, 6, 8, 11 and 13 semitones lie very close to a straight line. The mean responses for intervals of 2, 3, 5, 7, 10 and 12 semitones are significantly higher than the line (Sect. 5.1.1). This suggests that pitch commonality contributes to the pitch relationship between sequential tones spanning the latter intervals. The strongest effect occurs at P8 (the octave); P4 and P5 follow; and M2, m3, and m7 show small but significant effects. The effects at the M3 and M6 are insignificant, suggesting that the relationship between sequential tones an M3 or M6 apart in music is mainly learned (e.g. from regular exposure to major and minor triads).

The responses by musicians for complex tones are similar to those by non-musicians, except that they show an exaggerated positive emphasis on the octave and negative emphasis on the tritone, apparently due to interval recognition and cultural conditioning. The increased range of the confidence intervals for musicians is due to their smaller number (9, as against 21 non-musicians); it does not reflect any uncertainty in their responses.

Figure 5.5b compares similarity ratings of rising and falling intervals. (Note that all results include data from both musicians and non-musicians.) The octave effect for *complex* tones was significantly greater for *rising* intervals, as in the previous (piano tones) experiment (Sect. 5.4.3). Again, this may be understood in terms of the effect of context on the pitch ambiguity of a complex tone, taking into account that the main susidiary pitch of a complex tone in the central pitch range normally lies an octave above the main pitch.

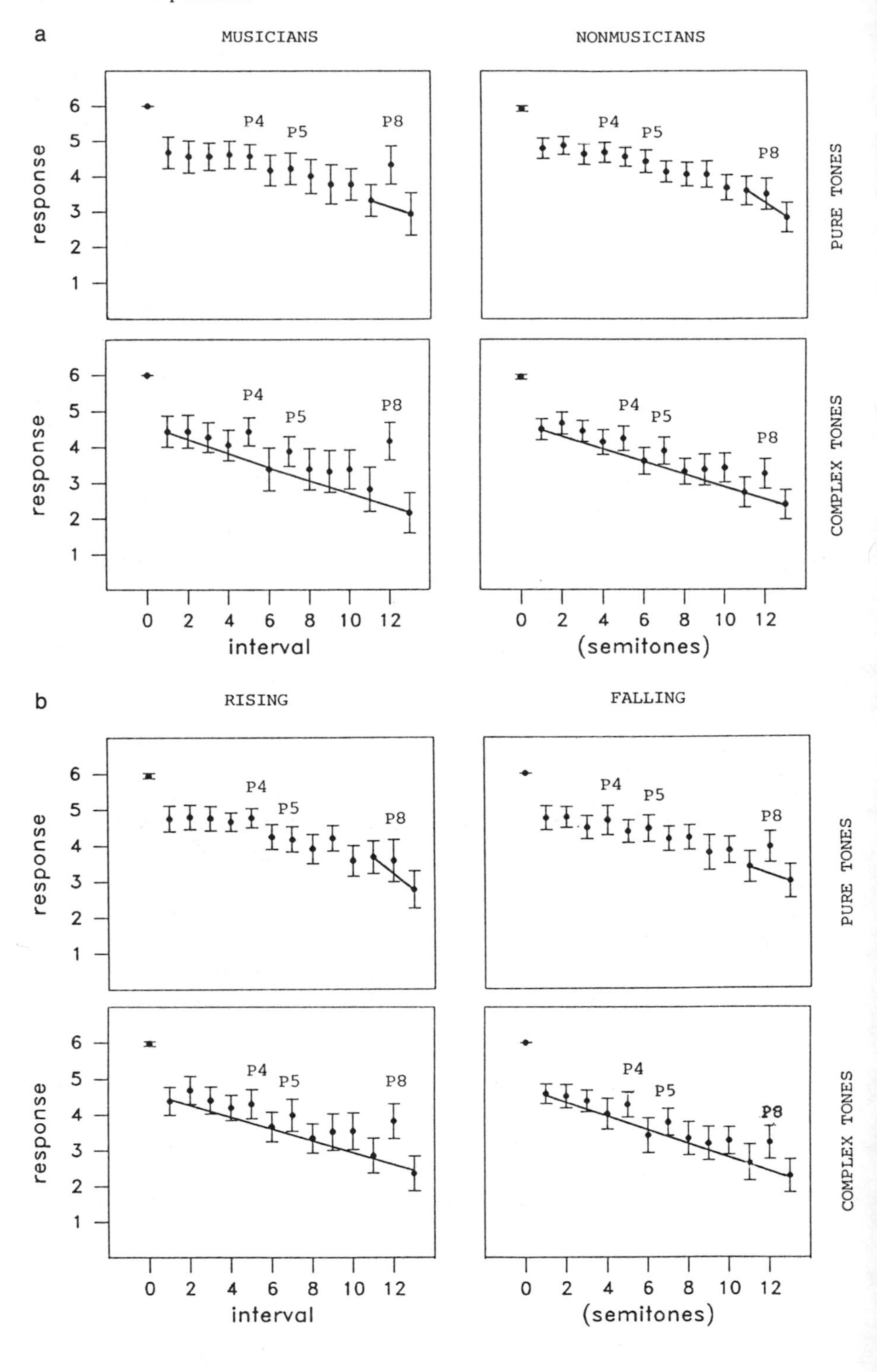
a
MUSICIANS
NONMUSICIANS
PURE TONES
COMPLEX TONES
response
P4
P5
P8
0 2 4 6 8 10 12
interval
(semitones)
b
RISING
FALLING
PURE TONES
COMPLEX TONES
response
P4
P5
P8
0 2 4 6 8 10 12
interval
(semitones)

The octave effect for *pure* tones was significantly greater for *falling* intervals than for rising intervals. This appears to be because the main subsidiary pitch of a pure tone lies an octave *below* rather than above its main pitch (Sect. 2.4.5).

5.5.4 Grouping

Listeners (9 musicians and 21 non-musicians) were classified into groups on the basis of their responses. Those whose responses to pairs of complex tones at intervals of 12 and 7 semitones exceeded responses to pairs at intervals of 11 and 6 semitones by an average of one response category or more were assigned to a *harmonic relationship* (*complex tones*) group; this group contained 5 musicians and 7 non-musicians. By an analogous method, 3 musicians and 2 non-musicians were assigned to a *harmonic relationship* (*pure tones*) group. Those 8 musicians and 14 non-musicians whose response to pairs of tones of similar timbre at intervals of 1 or 2 semitones exceeded responses for 13 or 10 semitones by an average of one response category or more were assigned to a *pitch distance* group. The 6 musicians and 19 non-musicians whose responses for pairs of the same pitch and similar timbre exceeded responses for pairs of the same pitch and different timbre by an average of one or more were assigned to a *timbral similarity* group.

The grouping analysis shows that timbral similarity and pitch proximity were generally more important than harmonic relationship, and that harmonic relationship between complex tones was generally more important than harmonic relationship between pure tones. The fact that only 2 of 21 non-musicians were assigned to the harmonic relationship (pure tones) group squares with the assumption that octaves were recognized only seldom by non-musicians.

5.5.5 Conclusions

Similarity ratings were mainly influenced by the sensory parameters timbral similarity and pitch distance. Harmonic effects were weaker, but still significant ($p < 0.05$) in many cases. The existence of both sensory and cultural contributions to harmonic aspects of the similarity of both pure and complex tones covering octave intervals was suggested.

The weakest harmonic effect was the sensory contribution of the similarity of pure tones an octave apart. It was only significant in the case of failing in-

Fig. 5.5 a, b. Results of similarity of synthetic tones experiment I. The 95% confidence intervals of mean results for pairs of pure tones (*upper panels*) and complex tones (*lower panels*); diagonal lines indicate the assumed pattern of results in the absence of harmonic effects (pitch commonality). **(a)** Results of 9 musicians, 2 data each (*left panels*) and 21 non-musicians, 2 data each (*right panels*). **(b)** Results of all 30 listeners for rising tone pairs (*left panels*) and falling tone pairs (*right panels*)

tervals. A possible explanation is that only the first in a sequential pair of pure tones has perceptible subharmonic pitches: the pitch of the second tone is disambiguated by the context created by the first. Similarly, the harmonic effects for complex tones on octave apart was significantly larger for rising intervals than for falling intervals, apparently because the first in a pair of complex tones has a more salient pitch at its upper octave than the second in the pair. This effect was found in the previous experiment (using piano tones) and confirmed in the present experiment (using synthetic tones). For complex tones, sensory harmonic effects (i.e. harmonic effects in the absence of musical interval recognition) were strongest at the octave, weaker at the fourth and fifth, and weaker still (but still significant) at the major second, minor third, major sixth and minor seventh.

The pattern of similarity ratings as a function of pitch distance was similar to ratings of familiarity and frequency of occurrence of melodic intervals [Jeffries 1972], and to rates of correct recognition of melodic intervals [Terhardt et al. 1986]. All these effects may be either sensory (due to pitch distance and pitch commonality) or cultural (due to the relative numbers of intervals in actual melodies [Jeffries 1974]). The cultural effect, if it is important, apparently depends on the sensory effect anyway, via music history and cultural conditioning (Sect. 3.1.1). It may therefore be described as "indirectly sensory" (Sect. 5.1.2).

5.6 Similarity of Synthetic Tones II

5.6.1 Introduction

In an experiment to see how well pure probe tones "fit in, musically" with a previously heard diatonic scale (also consisting of pure tones), Jordan [1987] obtained tone profiles with peaks corresponding to diatonic scale steps. The peaks had half-widths of about half a semitone. It may be concluded that sequential tones less than half a semitone apart in a musical context are assigned to the same pitch category. This is consistent with theory and data on the categorical perception of relative pitch [Burns and Ward 1978] (Sect. 2.5.3).

One of the aims of the present experiment was to test how categorical perception influences the results of similarity ratings. Some non-chromatic (microtonal) intervals were included, in order to investigate: how far apart pure tones must be before they are heard to be "different" rather than "the same" in an experiment of this kind; whether complex tones spanning non-chromatic intervals are less similar than complex tones spanning ordinary chromatic intervals; and whether there is a sensory basis for the similarity of complex tones at the major third (as no significant effect had been found for this interval in the previous experiment).

Previous experiments (Sects. 5.4, 5) tested the similarity only of pure and full complex tones; other tone types (square waves, residue tones) had been included only to encourage listeners to concentrate on timbral variations. In the present experiment, results were analyzed not only for pairs of ordinary pure and complex tones but also for pairs of residue and octave-spaced tones. Of interest were the following possibilities: that residue tones might show weaker harmonic effects than complex tones due to the physical absence of the fundamental; and that octave-spaced tones might show a harmonic effect at the interval of a fourth, due to subfifth tone sensations predicted by the model (Sect. 6.1.5).

Previous experiments had established in a qualitative way that pitch commonality influences similarity ratings of sequential tones. In the present experiment, this was tested quantitatively by comparing results with calculations according to a model of pitch relationship (Sect. 4.6).

5.6.2 Method

Listeners. 8 musicians and 12 non-musicians took part in the experiment (Sect. 5.1.2).

Equipment was the same as for the previous experiment (Sect. 5.5.2).

Sounds. Spectra of representative tones are shown in the upper panels of Figs. 5.6. Full complex tones had 16 harmonics with a spectral envelope gradient of -6 dB/octave (as in the previous experiment). The pure tone components of residue tones were confined to the range C_4 (middle C)$-C_7$; their amplitudes were constant in the range C_5-C_6, and fell linearly to zero in the sidebands C_5-C_4 and C_6-C_7. Octave-spaced tones contained octave-spaced pure tone components in the range C_2-C_8; their amplitudes were constant in the central range C_4-C_6 and fell linearly to zero in the sidebands C_4-C_2, and C_6-C_8. Waveforms were sampled at a rate of 18 kHz. Levels were adjusted for roughly equal loudness.

In all, 98 different tone pairs were presented in the experiment. No anchor pitches were used. Of the tone pairs, 22 consisted of pure tones in the range C_4-C_5 covering the intervals $\pm$(0, 0.25, 0.5, 0.75, 1, 1.25, 1.5, 1.75, 2, 11 and 12) semitones; 38 pairs consisted of full complex tones in the range A_2-C_4 covering intervals $\pm$(0, 0.5, 1, 1.5,..., 7, 7.5, 8, 11, 12) semitones; 14 pairs consisted or residue tones with (missing) fundamental frequencies in the range C_3-C_4 covering the intervals $\pm$(0, 1, 5, 6, 7, 11, 12) semitones; 24 pairs consisted of octave-spaced tones covering the intervals 0, 6, $\pm$(0.5, 1, 1.5,..., 5, 5.5) semitones.

Timbre (i.e. tone type) was the same within each trial. It was considered unnecessary to include additional trials with different timbre (as in previous experiments) as the experiment already contained a lot of timbral variation.

Procedure. Each trial consisted of a repeated tone pair. Each tone had a duration of 0.2 s. The time interval between tones in a pair was 0.2 s, and between pairs in a trial, 0.45 s. Listeners had as long as they wished to respond, but they were asked to do so quickly and spontaneously. They rated the

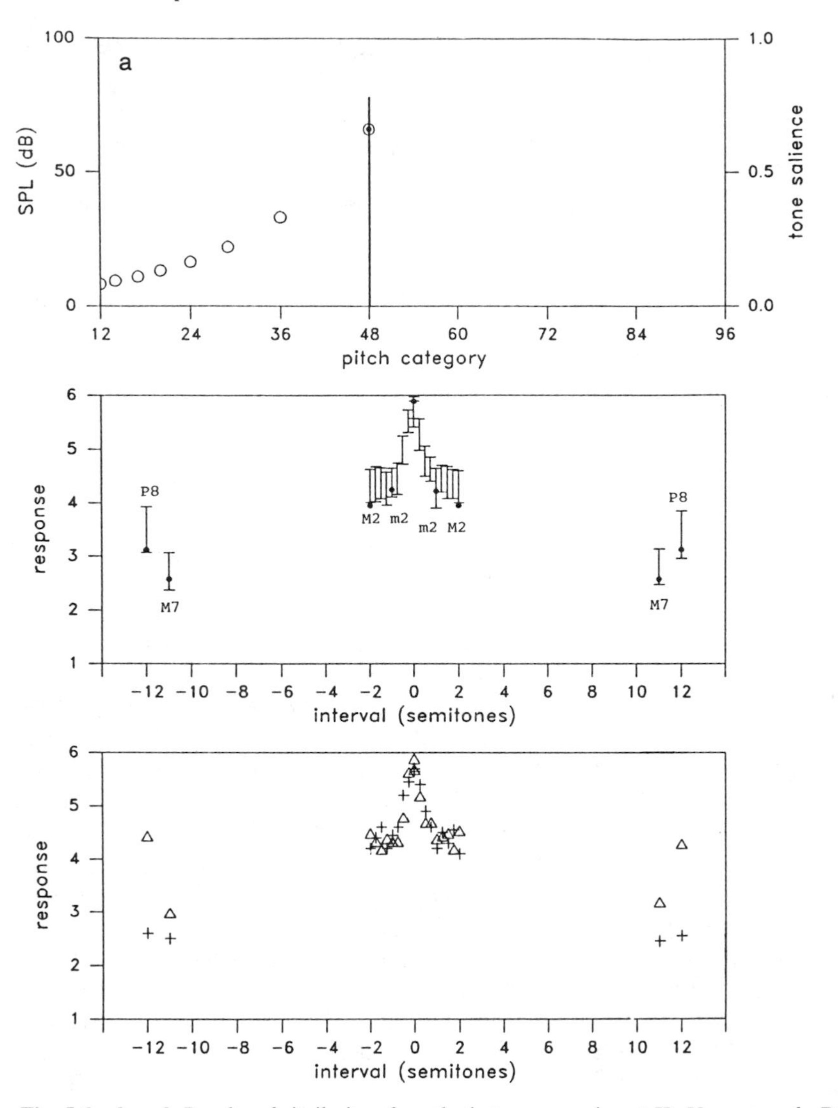

Fig. 5.6 a, b c, d. Results of similarity of synthetic tones experiment II. *Upper panels*: Pure tone components (*vertical lines*, *left scale*), pure tone sensations (*points*, *right scale*) and complex tone sensations (*circles*, *right scale*). *Centre panels*: 95% confidence intervals of mean responses of all 20 listeners, 2 runs each (*bars*); calculations according to (4.37) with $k_M = 20$, $k_T = 1.0$, $k_S = 0.6$, and $k_R = 0.25$. *Lower panels*: Mean responses of 10 OE listeners, 2 runs each (*triangles*), and 10 non-OE listeners, 2 runs each (*crosses*). (**a**) Pairs of pure tones

similarity of the tones on a 6-point scale. The response keyboard was labelled "very dissimilar" (at the left) and "very similar" (at the right).

Trials were presented in a random order, which differed for each run of the experiment. Each listener performed to experiment twice.

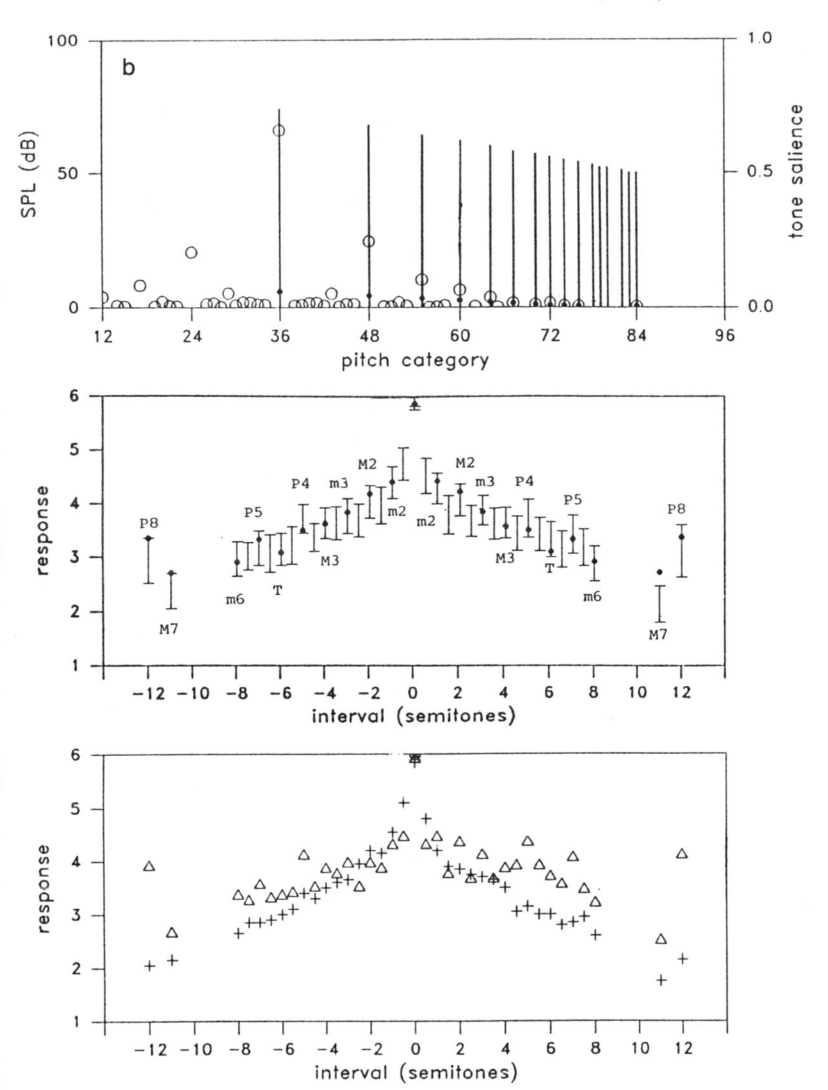

Fig. 5.6 *(continued).* **(b)** Pairs of full complex tones

5.6.3 Grouping and Results

The degree to which responses for octaves exceeded responses for major sevenths was calculated, taking into account responses for pure, full complex and residue tones. The 10 listeners who emphasized octaves to the greatest degree were assigned to an "octave equivalence" (OE) group. In the light of previous experiments, the ratio of musicians to non-musicians in this group was expected to be higher than in the other (non-OE) group. Surprisingly, these

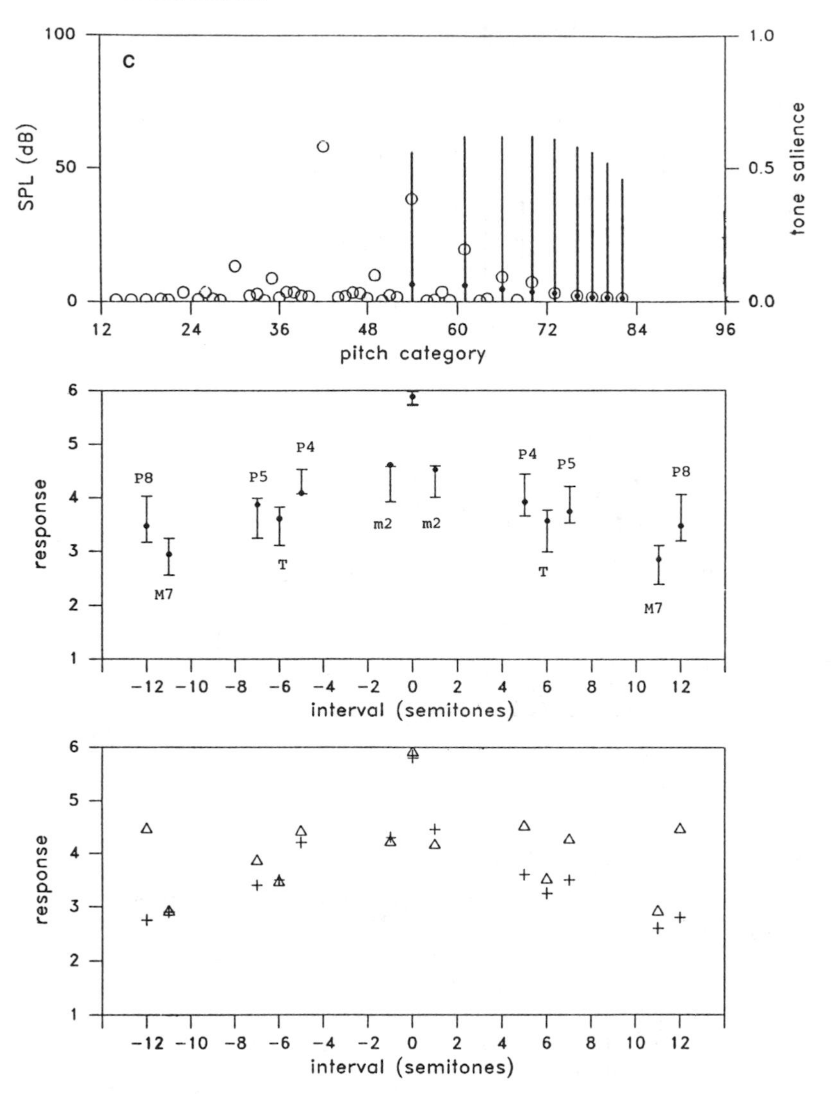

Fig. 5.6 *(continued).* **(c)** Pairs of residue tones

ratios turned out to be exactly the same: each group contained 4 musicians and 6 non-musicians.

Results appear in Fig. 5.6. The lower panels illustrate differences in response strategies by a comparison of the results of the OE and non-OE groups. In general, members of the non-OE group showed no harmonic effects at all. Some possible interpretations of this are (i) that the pitch commonality hypothesis is false, and harmonic effects are merely learned from musical experience; (ii) that the effect of pitch commonality is weak and may, in some circumstances, be completely dominated by the effect of pitch distance; and

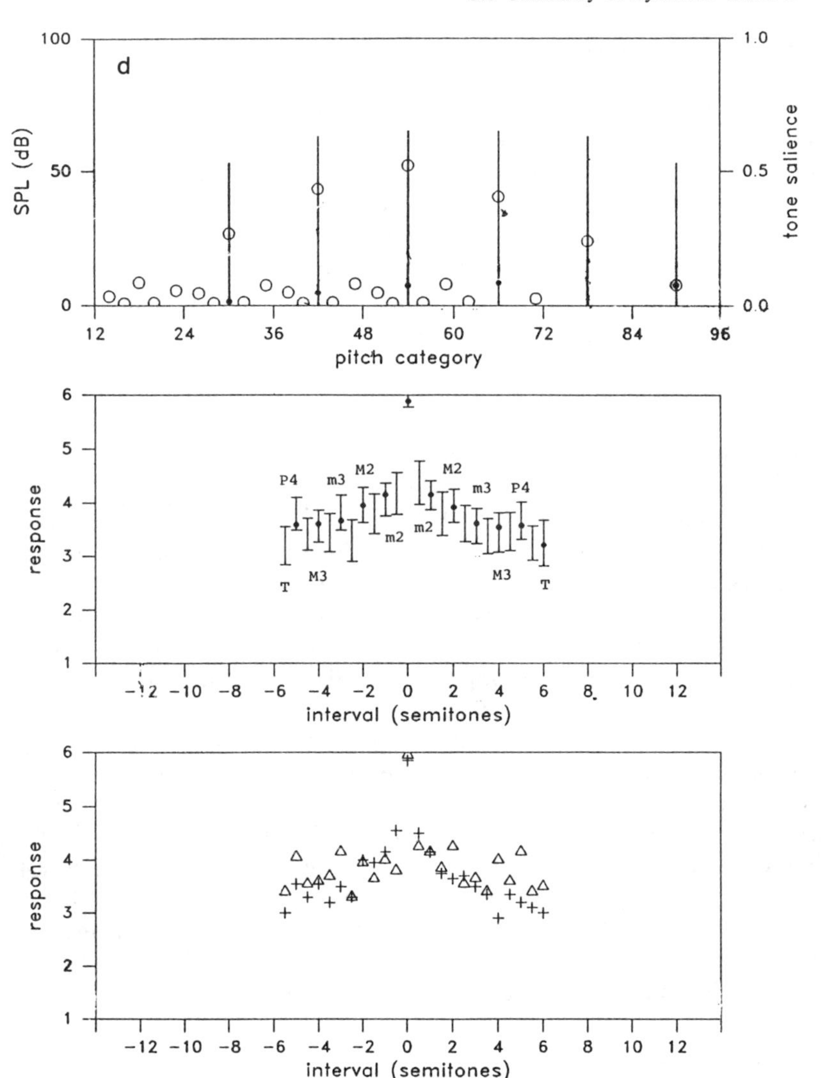

Fig. 5.6 *(continued).* **(d)** Pairs of octave-spaced tones

(iii) that the lack of harmonic effects among the non-OE listeners is a direct consequence of the procedure by which the group was selected. Evidence from other experiments in this study favours the second and third possibilities.

According to Fig. 5.6a, sequential pure tones of 0.2 s duration and spanning up to about 0.5 semitones were interpreted as identical, by both OE and non-OE listeners. This may be regarded as a kind of categorical perception (Sect. 2.5.3). The category width of 0.5 semitones is about ten times greater than the local difference threshold, see Fastl and Hesse [1984]. Possible reasons for this include (i) context: listeners became used to the relatively large pitch distances in the present experiment; and (ii) task: in the present experiment listeners were asked to rate similarity, not identity.

The overall results for pairs of complex tones (Fig. 5.6b) generally confirm previous conclusions on the sensory basis of harmonic relationships between sequential tones, although in several cases insufficient trials had been carried out to establish significance. Significant effects (Sect. 5.1.1) were found for the intervals M2, P4 and P8 for both rising and falling intervals, and for the m3 (rising) and P5 (falling). There was no significant effect at the M3.

Significant harmonic effects were found for residue tones at the intervals P4, P5, and P8 (Fig. 5.6c) in spite of their missing fundamentals. This may be understood in terms of Terhardt's theory of pitch, according to which a residue tone evokes tone sensations in essentially the same pitch pattern as a full complex tone.

The results for octave-spaced tones (Fig. 5.6d) include significant effects at rising and falling perfect fourths. These are consistent with the model's prediction that octave-spaced tones evoke subfifth tone sensations (upper panel). The lack of a significant effect at the major third suggests that subthird tone sensations of octave-spaced tones are imperceptible. The effect at the falling major second may be cultural, due to the pervasiveness of this progression (about a quarter of all intervals in popular melodies are falling M2s [Jeffries 1974]). Alternatively, it may be sensory, due to pitch commonality.

The effect at the falling minor third in Fig. 5.6d is unlikely to be sensory in origin. Given that the calculated pitch properties of octave-spaced tones (upper panel) are valid, such a sensory effect could only be due to the coincidence of (weak) subfifth pitches in the first tone of the trial with (very weak) subthird pitches in the second. The effect may be related to the importance of the minor third in children's songs, at least in the West; but exactly how is not clear. Critical bandwidth (Sect. 4.3.1) equals three semitones in the most important frequency region for spectral pitch (the pitch dominance region), however, this is unlikely to be relevant, as it only affects simultaneous sounds.

5.6.4 Modelling

Equations (4.12–37) were used to model responses for those 58 pairs of tones which spanned chromatic intervals (i.e. whole numbers of semitones). Values of the free parameters producing optimal fit between calculations and responses were $k_M = 20$, $k_T < 1$, $k_S = 0.6$ and $k_R = 0.25$. Of these, the values for k_M, k_S and k_R are fairly typical. The low value of k_T (<1, instead of about 3) may be due to effects of musical conditioning (e.g. interval recognition) not accounted for by the model.

The difference between the optimal value of k_R for the OE group (0.42) and for the non-OE group (0.10) was a direct consequence of the way the two groups were selected. Otherwise, parameter values were much the same for the two groups. Calculations fit mean responses more closely for the non-OE group than for the OE group, suggesting that members of the OE group were influenced more than members of the non-OE group by cultural effects not accounted for in the model.

The same model simulated results for pairs of pure, full complex, residue and octave-spaced tones. Calculations agreed with mean responses over all listeners in all but 5 of the 58 modelled trials. Discrepancies were expected on the basis of statistical fluctuations in 5% of 58, or 3 trials. The success of the modelling procedure in this experiment was partly due to the relatively large spread in the results: the results would have improved, and the number of discrepancies between calculations and mean responses (presumably) increased, had there been more listeners.

5.6.5 Conclusions

Despite the mixture of sensory and cultural effects contributing to the results, it appears that the sensory contribution to the harmonic relationship perceived between sequential tones was lower for non-chromatic intervals than for their chromatic neighbours. Most listeners heard pure and complex tones a half a semitone or more apart to be different, implying that the pitch commonality of tones at intervals mistuned by half a semitone is lower than the pitch commonality of well-tuned intervals. (In actual music, the flow of perceptual information can be much higher than it was in the present experiment, and larger mistunings between sequential tones can go unnoticed, see Burns and Ward [1978].)

5.7 Similarity of Chords

5.7.1 Introduction

The algorithm used to model the results of the previous experiment (Sect. 4.6) quantitatively predicts the strength of harmonic and voice-leading relationships between musical chords (Sects. 3.2.3, 6.2.1, 2). The present experiment aimed to test the accuracy of the model in the case of pairs of major triads in close position, by means of similarity ratings. It was assumed that similarity ratings of musical chords out of context would reflect the sensory basis of chord relationships in context.

Krumhansl et al. [1982] presented listeners with pairs of chords made up of octave-spaced tones, and asked them to rate on a 7-point scale "how well the second chord followed the first". In their multidimensional scaling solution of the results, pairs of chords in strong harmonic and melodic relationships and of similar consonance tended to be close together. Examination of the raw data (kindly supplied by Dr. Bharucha) showed the presence of an additional strong effect: dissonant chords followed by consonant chords were given high ratings (i.e. "followed well"). For example, apparent similarity was enhanced if a diminished triad resolved satisfactorily onto a major or minor triad.

The present experiment aimed to investigate the contributions only of harmonic relationship (i.e. pitch commonality) and pitch distance to pitch relationships between musical chords. The effect of resolution was avoided by holding consonance relatively constant (by using only major triads).

5.7.2 Method

Listeners and **equipment** were the same as for the multiplicity and pitch analysis experiments (Sect. 5.2.2).

Sounds. These are notated musically in Fig. 5.7. They include all major triads playable using the twelve tones $C_4 - B_4$ inclusive. The frequency spectrum of chord 1 was the same as that of sound 1 in the multiplicity and pitch analysis experiments (upper panel of Fig. 5.3a). The complex tones from which the chords were built had roughly equal loudness when heard separately. Each chord lasted 0.2 s. Amplitude envelopes were shaped to remove onset of offset clicks.

Fig. 5.7. Music notation of sounds presented in the similarity of chords experiment

Procedure. Each trial consisted of a repeated pair of chords. Each chord lasted 0.2 s. The time interval between chords was 0.3 s, and between repetitions of pairs of chords, 0.6 s. The spectra of the chords in each trial were shifted through in randomly chosen distance of -2 to $+2$ semitones. The shifts, which were determined separately for each trial and each run, were intended to confuse any listeners with perfect pitch.

Listeners were allowed as long as they wished to respond, but were asked to do so quickly and spontaneously. They rated each pair of chords by pressing one of four buttons labelled "very dissimilar", "sort of dissimilar", "sort of similar", and "very similar". Responses were recorded as numbers 1–4 (respectively).

The experiment comprised 132 trials: each of 12 chords was compared with each other chord (in two orders of presentation). The trials were presented in a random order which was different for each experimental run. Listeners took a break after the 66th trial.

5.7.3 Results

Mean responses for pairs of chords in which the first chord was chord 1 (the C major triad) are graphed in Fig. 5.8. In general, responses were higher for

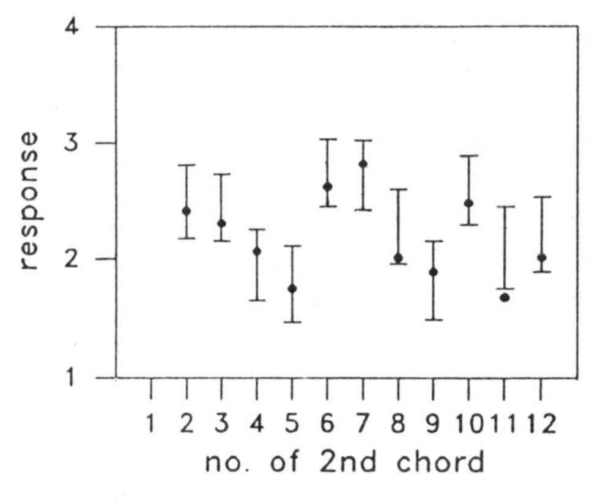

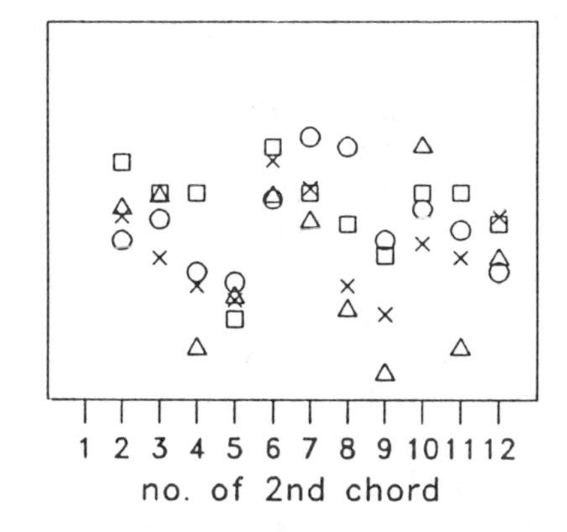

Fig. 5.8. Results of the similarity of chords experiment for pairs in which the first chord was chord 1. *Left panel*: The 95% confidence intervals of mean responses of all 39 listeners (*bars*); calculations according to (4.37) with $k_M = 28$, $k_T = 3.5$, $k_S = 0.3$, and $k_R = 0.3$. *Right panel*: Mean results of 10 Western musicians (*triangles*), 8 Eastern musicians (*squares*), 9 Western non-musicians (*crosses*) and 12 Western children (*circles*)

chords whose voices led well according to music theory; for chords which had a tone in common (N.B. no chords had more than one tone in common); and for chords whose roots were close on the cycle of fifths. In the figure, responses are highest when chord 1 (C) is compared with chords 6 (A*b*), 7 (A) and 10 (F). These are the only chord relationships in the set combining good voice-leading and a tone in common. Responses are lowest when the C chord is compared with chords 4 (E*b*), 5 (E) and 9 (B). Voice-leading for these progressions is weak. Voice-leading in the progression from C (1) to G (12) is also weak, but the effect of this appears to be offset by closeness on the cycle of fifths.

5.7.4 Modelling

Calculations of the model (4.37) were fitted to mean responses to all 132 trials for each group of listeners. Values of the free parameters producing optimal fit between calculations and responses were $k_M = 28$, $k_T = 3.5$, $k_S = 0.3$ and $k_R = 0.3$. Values of the parameters k_M, k_T and k_R were consistent with previous findings. The simultaneity perception parameter k_S was lower than its typical value of 0.5, apparently because of the relatively holistic orientation of the experimental task.

The optimal value of the pitch relationship perception parameter k_R was highest, as usual, for the Western musicians (0.55), indicating that they were influenced by pitch commonality more than were the other listeners. The Western musicians may have been responding to a large extent on the basis of harmonic relationships learned from music. The optimal value of k_R was lowest (0) for the Eastern musicians, for whom pitch distance was more important than pitch commonality. Perhaps these listeners tended to hear chords as single, non-harmonic bell-like sounds, due to their familiarity with *gamelan* sounds. k_R was also fairly low for the Western non-musicians and children (0.25): "non-musicians seem to be most influenced by considerations of pitch height and not by the complex hierarchical and ratio relations among pitches that a learned tonal system engenders" [Monahan et al. 1987, p. 599, in reference to Krumhansl and Kessler 1982]. Optimal values of the other free parameters (k_M, k_T and k_S) varied relatively little among different groups.

Calculated responses disagreed with mean responses in 17 of 132 cases. This exceeds the expected number of about 7 (5% of 132, due to the statistical spread about 95% confidence intervals). Clearly, the model does not account for all experimental effects. Most of the discrepancies were found to be explicable in terms of familiarity with music; for example, pairs of major triads whose roots are a fifth apart are familiar (e.g. from perfect cadences) while pairs whose roots are a minor third apart are unfamiliar.

5.7.5 Conclusions

From a music-theoretical perspective, similarity of musical chords was explicable in terms of voice-leading, tone commonality and the cycle of fifths. The sensory basis for voice-leading was assumed to be pitch distance; for tone commonality and closeness on the cycle of fifths, pitch commonality. A model based on pitch distance and pitch commonality (in the absence of musical experience) successfully modelled the results of the experiment.

Consistent discrepancies between calculations and results were mainly due to inadequacies of the model and to musical conditioning. It was assumed in the model that the only familiar perceptual pattern influencing the perception of musical chords is the pattern of pure tone sensations in a typical complex tone (e.g. a speech vowel). The model does not account for familiarity with patterns of complex tone sensations in musical chords, or with particular chord progressions.

5.8 Discussion

5.8.1 Modelling

Values of parameters producing best fit between calculations and mean responses are summarized in Table 5.1. The table shows that parameter values varied considerably, depending on the type of sounds presented in an experiment and the task performed by the listeners. For example, the tone perception parameter k_T was unusually high (5.6) for the multiplicity experiment, apparently because the manner of presentation of sounds and the experimental task encouraged the hearing out of pure tone components; and the pitch relationship perception parameter k_R was especially high (0.65) for the pitch analysis experiment, as the task explicitly involved pitch commonality rather

Table 5.1. Modelling results of experiments

Experiment	N_t	N_d	k_M	k_T	k_S	k_R
Multiplicity	10	78	29	5.6	0.51	–
Pitch analysis	158	39	45	3.5	0.3	0.65
Similarity of tones	58	40	20	<1	0.6	0.25
Similarity of chords	132	39	28	3.5	0.3	0.3
Music theory values			25	3	0.5	–

N_t: Number of trials
N_d: Number of data per trial
k_M: Masking parameter (4.14)
k_T: Tone (pure/complex) perception parameter (4.19)
k_S: Simultaneity (multiplicity) perception parameter (4.24)
k_R: Relationship (commonality/proximity) perception parameter (4.34)

than pitch distance. Extreme values of individual parameters in the table (e.g. $k_M = 45$ for the pitch analysis experiment; $k_T < 1$ for similarity of synthetic tones II) are hard to interpret in terms of listeners' response attitudes and strategies. Perhaps these values compensated somehow for the failure of the model to account for cultural effects. The "music theory values" in the table are the values of the free parameters chosen for music-theoretical use (Chap. 6).

The use of no less than four free parameters in the model may seem somewhat extravagant: *any* experimental results may be modelled successfully if enough free parameters are introduced. However, the minimum number of data points which may be *exactly* fitted using four free parameters is only six, if two degrees of freedom are counted for the linear regression between calculations and measurements. Four free parameters may successfully fit more than six random points if limited deviations between experimental and calculated values are allowed. The allowable deviations in the experiments of this study – the 95% confidence intervals – were relatively large, but the number of points to be fit simultaneously in each experiment was also high: the three experiments modelled by adjusting four free parameters included 158, 98 and 132 data points respectively. In the multiplicity experiment (Sect. 5.2.4), three free parameters were adjusted to produce a rather bad fit to only ten data points. It would be useful in future work to repeat this experiment, using more varied sounds, in order to test more thoroughly (and possibly improve) the multiplicity algorithm (4.23, 24).

5.8.2 Musical Universals?

The study would have benefited greatly from the introduction of a proper control group into the experiments – a group of listeners with minimal experience of Western music. The attempt to introduce such a group (the Indonesian musicians) was largely unsuccessful, because of their small number, and because the effect of exposure to Western music on their responses was uncertain.

Would it be possible for a group of people with negligible experience of Western music to participate meaningfully in experiments of the kind described here? The experimental philosophy of a study such as this is itself culture-specific: it is biased in favour of an analytical mode of solving problems, as well as of listening to and performing music. This way of thinking and making music would make little sense in some non-Western societies. Further, instructions about Western-style experimental tasks may be difficult to translate into languages of remote cultures. Translations, where possible, would be marred by problems of interpretation, including interpretation of the nature of music itself. For any of these reasons, it may be meaningless to try to compare experimental results of remote cultural groups.

I attempted in this study to separate sensory and cultural effects in music perception by comparing experimental results with the calculations of a sensory model. Another approach might be to compare the musical styles of two originally independent cultures. Ideally, such a study would be conducted independently of the music theories of the two cultures, as music theories are normally quite ethnocentric. The researcher would need to become deeply involved in the music of both cultures, and at the same time maintain scientific objectivity; and any objective methods of analysis (including quantitative models) would need to include a balance of elements from the different cultures.

In any case, similarities in musical styles, or in experimental results of people from different cultural backgrounds, do not necessarily indicate perceptual universals. For example, octave relationships in Western music have both sensory and cultural components (Sect. 3.3.1): the similarity of tones an octave apart is exaggerated by musical conditioning (see the responses of the musicians in Fig. 5.5a). Presumably, this is also true in other musical cultures in which octaves are important. Thus, exaggerated estimates of octave similarity by people from different musical cultures reflect (non-universal) similarities in their musical cultures, not (universal) sensory effects.

6. Applications

The model described in Chap. 4 predicts the number of audible harmonics in complex tones, the multiplicity and tonalness (and hence consonance) of musical tone simultaneities (tones, dyads and chords), the various possible pitches of simultaneities and their perceptual importances (saliences), and the roots of chords. It also predicts the strength of harmonic and melodic relationships between sequential musical sound (pitch commonality and pitch proximity). It quantifies sensory and cultural aspects of the tonality of chord progressions: repetition, sequential harmonic relationships (including the roots of broken chords), consonance, and implication (of triads, scales and keys). It enables "objective" psychoacoustical analysis of harmonic progressions, and may be applied in composition.

6.1 Simultaneities

6.1.1 Masking

The number of audible harmonics in a complex tone is limited by masking (Sect. 4.3). According to Plomp [1964], Marsenne, in 1636, reported hearing out 7 harmonics from a complex tone; Helmholtz [1863] reported hearing between 12 and 16; and Brandt, in 1861, heard 8 harmonics in gut strings and 13 in brass strings. Schouten [1940] reported 12 audible harmonics. Schenker [1906], in his discussion of the origins of harmony, assumed that only 6 harmonics were clearly audible. These differences are due to variations in physical spectra, in theoretical approach, and in the ability to listen analytically.

Plomp [1964] performed experiments to determine the number of directly audible harmonics in complex tones comprising harmonics of equal sound pressure level (SPL). He found that listeners could reliably distinguish between 5 and 8 harmonics, depending on the fundamental frequency of the tone. According to the masking algorithm of Terhardt [1979a] and of Terhardt et al. [1982b], up to 11 harmonics may be audible in such sounds. Equations (4.12–16) of the present model (Chap. 4) make similar predictions in the case of "typical" musical tones (4.8). These results and calculations are compared with each other in Table 6.1.

Plomp's experimental values are lower than Terhardt's theoretical ones. This is to be expected, as the degree to which a listener can ignore complex tone

Table 6.1. Number of audible harmonics in typical full complex tones

Note	C_1	C_2	C_3	C_4	C_5	C_6	C_7	C_8	C_9
P	12	24	36	48	60	72	84	96	108
N_1		7	8	7	6	5	5		
N_2	1	10	11	9	7				
N_3	8	10	11	10	10	10	7	3	1

P: Pitch category
N_1: Measured number of distinguishable harmonics in complex tones of loudness 60 phon and with 12 harmonics of equal SPL [Plomp 1964]
N_2: Calculated number of audible harmonics in complex tones with many harmonics of 60 dB SPL each [after Terhardt 1979]
N_3: Calculated number of audible harmonics, according to (4.16), of full complex tones [spectra specified by (4.8)]

sensations and concentrate only on pure tone sensations by analytical listening is limited. But Terhardt's calculated values lie near the upper ends of Plomp's error bars, indicating that Plomp's listeners sometimes heard all the harmonics Terhardt calculated to be audible.

The masking algorithm described in Sect. 4.3 differs from that of Terhardt [1979a] in three main ways.

(i) Specified values of critical bandwidth at low frequencies in Terhardt's algorithm are too high (Sect. 4.3.1), so masking at low frequencies is overestimated. According to his algorithm, for example, a full complex tone with fundamental frequency 33 Hz (C_1) and harmonics of equal amplitude has only one audible component (the fundamental). Specified critical bandwidths in the present model (4.12) are smaller, but not as small as some published measurements of critical bandwidth (Table 4.1).
(ii) The masking pattern of a pure tone is assumed in Terhardt's model to be triangular when SPL is plotted against pure tone height (critical-band rate). In the present model, the masking algorithm is simplified by assuming that the pattern is triangular when auditory level (level above the threshold of audibility) is plotted against pure tone height.
(iii) The high-frequency side of the masking pattern of a pure tone is normally less steep (in decibels per critical band) than the low-frequency side. This means that low-frequency pure tone components mask high-frequency components more than the other way around. For simplicity, this effect is neglected in the present model, and the magnitudes of the two gradients are set equal (4.14).

It was considered preferable in designing the current model to underestimate masking rather than to overestimate it. If masking is underestimated, sensitivity of calculated pitch properties to changes in overall level is reduced. This makes the model more suitable for music-theoretical applications, in which the overall level (loudness) of a chord is normally unimportant (Sect. 2.2.1). Underestimation of masking also allows for the possibility that slightly subthreshold tone components contribute to the formation of complex tone sensations: residue pitches can sometimes be heard in sounds containing only high, adjacent harmonics, which (due to masking) cannot be heard out separately [Moore and Rosen 1979].

According to the present model, isolated musical tones in the central musical pitch range have ten audible harmonics. Similarly, the harmonic template for the simulation of complex tone perception (Sect. 4.4.1) is limited to ten components. This is consistent with the theory that the template is acquired from experience of complex tones (Sect. 2.4.3).

6.1.2 Spectral Dominance

Spectral dominance [Fletcher and Galt 1950; Ritsma 1967] is an effect by which the precise pitch of a complex tone is most likely to be determined by tone components in a particular range of frequencies. Terhardt's model [Terhardt et al. 1982b] accounts for this effect explicitly by means of a *spectral frequency weight* function with a broad maximum at 700 Hz (F_5). Various versions of this function were tried out in the current model, but none significantly improved the correlation between calculations and the results of the pitch analysis experiment (Sect. 5.3), an experiment designed to be sensitive to effects of spectral dominance on tone saliences. The reason for this is not clear. It could involve the limited frequency range of the sounds used in the experiment, problems with intonation of the probe tones, or inadequacy of the simplified masking algorithm. Whatever the reason, it was clear that the effect of spectral dominance on the salience of pure and complex tone sensations was relatively small, and could safely be neglected for music-theoretical purposes.

The threshold of audibility limits the auditory levels of pure tone components of very low and very high frequency in a way roughly similar to the way in which the spectral frequency weight function modifies the saliences of corresponding pure tone sensations. The threshold of audibility for pure tones [Terhardt 1979a] falls below 10 dB SPL in the range C_4-C_9 (260 Hz to 8.4 kHz); and Terhardt's spectral frequency weight function exceeds 0.8 in the range C_4-C_7 (260 Hz to 2.1 kHz). By neglecting spectral dominance, the current model may place undue emphasis on pure tone sensations in the range C_7-C_9 (one octave either side of the top note of the piano). This range was of negligible importance in the pitch analysis experiment (Fig. 5.3). Some compensation for this effect is provided in musical applications of the model by the falloff in specified auditory levels of pure tone components at high frequencies and high harmonic numbers (4.8).

The most salient tone sensation evoked by a complex tone with a fundamental frequency above about C_6 (1 kHz) is normally the pure tone sensation at the fundamental: the upper components of such a tone do not contribute to its pitch [Flanagan and Guttman 1960]. This is accounted for in

Terhardt's model by a *fundamental frequency weight.* It, too, is neglected in the present model. As a result, the model predicts that the most salient tone sensations of "typical" musical tones in the range $C_6 - C_8$ (the top two octaves of the piano) are complex tone sensations when, in fact, they are pure tone sensations. This is not a serious drawback, as the difference between pure and complex tone sensations is not directly perceptible, and does not directly affect pitch relationships (pitch commonality and pitch distance).

6.1.3 Multiplicity

Calculated multiplicities (Sect. 4.5.1) of selected musical tone simultaneities are shown in Fig. 6.1.

The example of fourth-spaced chords in part (a) of the figure is appropriate: the fourth is a fairly typically sized interval, and fourth-spaced chords spanning less than five octaves contain no octave doublings (which tend to reduce multiplicity). Calculated multiplicity exceeds the number of notes for single notes and dyads. Here, the variable M is best interpreted as a measure of *pitch ambiguity.* The pitch ambiguity of a single complex tone is about two: its pitch may be heard either at the fundamental or at a number of other pitches (especially the upper and lower octaves; see Fig. 6.3 and Terhardt et al. [1986]). In unaccompanied melody and two-part counterpoint, context and streaming (coherent amplitude or frequency modulation of harmonics: [McAdams 1984]) generally indicate the actual number of voices. In three-part music, multiplicity corresponds closely to the actual number of voices, making three perhaps the ideal number of voices for a contrapuntal passage in which all voices are important. Counterpoint in four or more voices is harder to write or listen to, as it is hard to follow (segregate) more than two or three voices

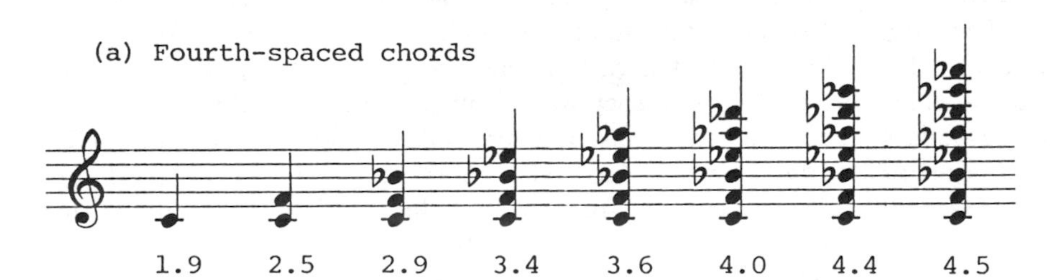

Fig. 6.1. Calculated multiplicity M, (4.24), of simultaneities of full complex tones (4.7, 8)

at once [Bregman and Campbell 1971]. For this reason, only two or three of the voices in a contrapuntal passage such as a Bach fugue are normally active at a given time; the others fill in the harmony.

Of course there is a big difference between picking out the voices of a chord and picking out the melodies of a contrapuntal passage. In the case of a progression, the task is made easier by streaming, i.e. by the horizontal context of each tone sensation. But it is hard to analyze and understand the structure of two melodies simultaneously [Sloboda 1985, p. 167]. The calculated multiplicity of a chord gives only a rough (and relative) indication of the number of separately perceptible melodies to which its tones may belong.

The musical examples in part (b) of the figure show the dependency of multiplicity on the intervals between notes in dyads and chords. Dyads typically have calculated multiplicities in the range 2.1 – 2.7, close-position triads in the range 2.8 – 3.0, and sevenths in the range 3.2 – 3.7. More consonant combinations of notes tend to produce lower calculated multiplicities, as they blend or "fuse" more easily.

6.1.4 Tonalness

The tonalness of a sound depends on the audibility of its (pure/complex) tone components (Sect. 3.2.2). *Pure tonalness* depends on the number and audibility of a sound's pure tone components; *complex tonalness*, on the audibility of a sound's most audible complex tone component. The two parameters are defined quantitatively in Sect. 4.4.3 in such a way that they both equal one for a "typical" harmonic complex tone at middle C. In most musical applications, the parameters are closely correlated with each other, so that references to "calculated tonalness values" in the following may be regarded as referring to either pure or complex tonalness, or both.

Calculated tonalness values for single musical tones in different registers are shown in Fig. 6.2a. Each tone has 16 harmonics with "typical" auditory levels (4.8). Middle C (C_4) has the highest calculated tonalness. Tones in the middle of the pitch range of music (and speech) have higher tonalness than very high or very low tones. In general, tones with higher tonalness are more common in music than tones with lower tonalness.

Calculated tonalness values for close position triads (Fig. 6.2b) are highest in registers 4 and 5. Values drop sharply in low registers due to masking, which reduces the audibilities of pure tone components. According to the model, the triad in register 1 (the first one analyzed) is almost completely pitchless: only a few pure tone sensations (corresponding to upper harmonics of the tones of the chord) are (barely) audible. This may be confirmed by playing the triad on the piano.

Calculated tonalnesses of dyads (simultaneous tone pairs) are set out in Fig. 6.2c. Calculated roughnesses [Hutchinson and Knopoff 1978] are also listed. The octave has easily the highest calculated tonalness – physically and perceptually, it is almost the same as a single complex tone. The fifth has the

(a) Single tones in various registers

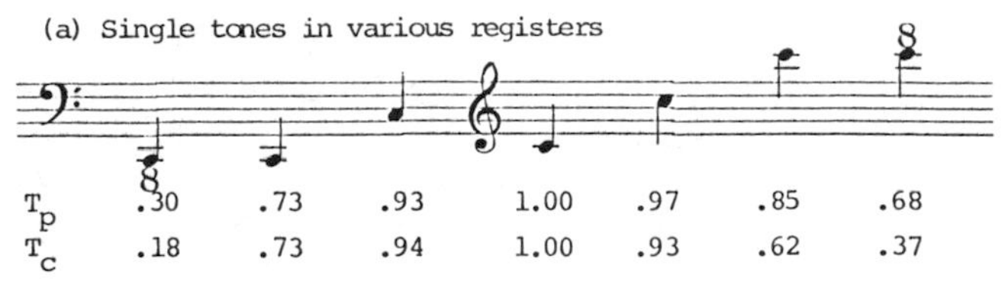

T_p	.30	.73	.93	1.00	.97	.85	.68
T_c	.18	.73	.94	1.00	.93	.62	.37

(b) Triads in various registers

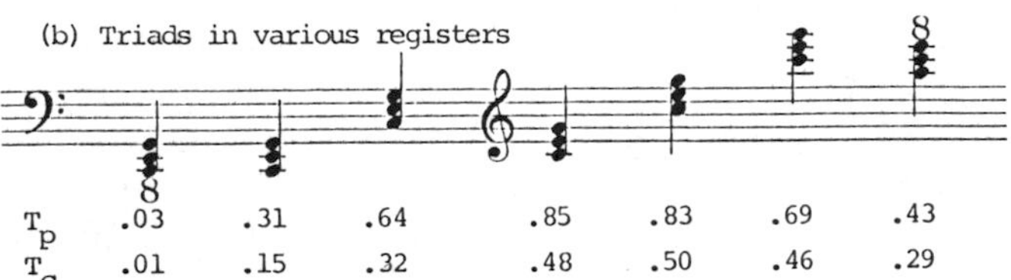

T_p	.03	.31	.64	.85	.83	.69	.43
T_c	.01	.15	.32	.48	.50	.46	.29

(c) Dyads in middle register

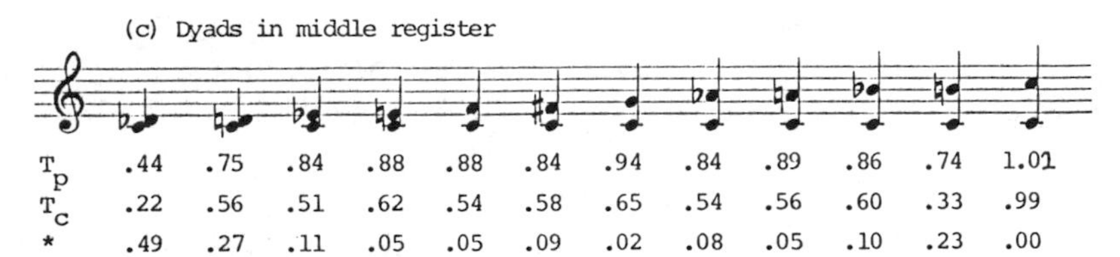

T_p	.44	.75	.84	.88	.88	.84	.94	.84	.89	.86	.74	1.01
T_c	.22	.56	.51	.62	.54	.58	.65	.54	.56	.60	.33	.99
*	.49	.27	.11	.05	.05	.09	.02	.08	.05	.10	.23	.00

(d) Triads in middle register

T_p	.85	.77	.84	.83	.82	.79	.72	.76	.79	.79
T_c	.48	.38	.43	.42	.38	.32	.35	.40	.37	.44
$\overline{T}_p$		.82			.81			.76		
$\overline{T}_c$		.43			.37			.37		

(e) Seventh chords in middle register

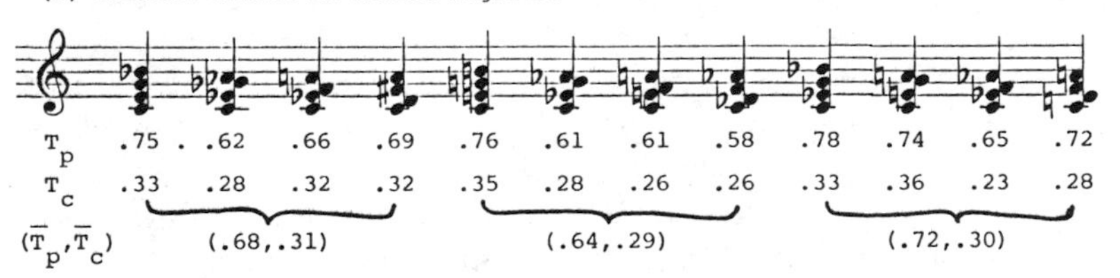

T_p	.75	.62	.66	.69	.76	.61	.61	.58	.78	.74	.65	.72
T_c	.33	.28	.32	.32	.35	.28	.26	.26	.33	.36	.23	.28
$(\overline{T}_p, \overline{T}_c)$		(.68,.31)				(.64,.29)				(.72,.30)		

(f) C major sixth chords in various spacings

T_p	.74	.78	.79	.80	.47	.55	.58	.63
T_c	.36	.34	.36	.23	.23	.18	.28	.27

(g) Triads in various doublings

T_p	.85	.79	.88	.84	.84	.82	.73	.78	.80
T_c	.63	.40	.53	.47	.51	.35	.40	.53	.50

second highest value. The minor second has the lowest tonalness (and the highest roughness, see Sect. 3.2.2). The major second and minor seventh intervals also have quite low tonalness (and high roughness) values. The other intervals vary relatively little with respect to their calculated tonalness. This suggests that, in two-part writing where homogeneity of consonance is desirable, all intervals beside the seconds, the major seventh and the octave may be freely used, including the tritone.

Malmberg [1918] investigated the consonance of dyads such as those in Fig. 6.2c by presenting them to and discussing them with a panel of eight academics (five musicians and three psychologists). The "blending", "smoothness", "fusion" and "purity" of dyads, produced by both tuning forks and the piano, were assessed. The rank order obtained after averaging over these various conditions, from least to most consonant, was m2, M7, M2, m7, T, m3, m6, P4, M3, M6, P5, P8. This order is similar to the order of calculated tonalnesses and roughnesses of the intervals in the figure. It may be concluded that the rank order of consonance of dyads of complex tones within the octave is relatively stable with respect to the way in which consonance is defined (blending, tonalness, etc.) and typical changes in the spectral content of the tones making up the dyads (Malmberg did not specify the spectral composition of the tones used in his experiment).

Figure 6.2d shows calculated tonalnesses for various close position triads in register 4. Values vary relatively little over this set, in accordance with the small range of consonance ratings of these chords given by untrained listeners [Roberts 1986]. The values nevertheless show some consistent and musically sensible trends. Triads including perfect fifth or fourth intervals have higher pure tonalness than triads including tritones or minor sixths (augmented fifths). Major triads have higher complex tonalness than minor triads. Root position major and minor triads usually have higher tonalness than their inversions; the reverse is the case for the diminished triad.

The dissonance of the augmented triad in music theory is not reflected by its calculated tonalness; it appears to have cultural rather than sensory origins. The triad does not fit into "overlearned" diatonic scale patterns (Sect. 3.4.3); it is not part of a commonly used tetrad (whereas the diminished triad is part of the dominant seventh chord); and its root, and therefore its harmonic function in context, is highly ambiguous. Consequently, the triad is relatively uncommon, and unfamiliar. The role of roughness in this argument could be investigated by applying the model of Aures [1984], possibly in an appropriately simplified form.

Figure 6.2e presents calculated tonalnesses of common tetrads (seventh chords). Overall, tetrads are less tonal than triads. Values for different tetrads, like the values for different triads above, cover only a small range. The tetrad

---

Fig. 6.2. Calculated pure tonalness T_p, (4.21), and complex tonalness T_c, (4.22), of simultaneities of full complex tones (4.7, 8). *: Roughness [after Hutchinson and Knopoff 1978]. $\bar{T}_p$, $\bar{T}_c$: Means over groups of chords

with the highest pure tonalness is the minor seventh chord in root position, and that with the highest complex tonalness, the minor seventh in first inversion (or major sixth in root position). With the possible exception of the minor seventh, root position seventh chords are more tonal than their inversions. The relatively low calculated tonalness of the major-minor sevenths (the first four chords in the set) is due to the tritone interval between third and seventh [Hindemith 1940]. The major-minor seventh's importance in Western music appears to be due to its relative lack of root ambiguity (all notes correspond to harmonics of the root) and its key-defining property (because it includes the interval of a tritone, it is diatonic – dominant, in fact – in only one key [Browne 1981]).

The calculated tonalness values listed in Fig. 6.2 f are consistent with music-theoretical conventions (Sect. 1.1.2) that close spacing should be avoided in low and wide spacing in high registers (cf. also Fig. 6.2 b). Other variations in calculated tonalness in this set depend in a less straightforward way on the intervals between the notes in the chords.

The effect on calculated tonalness of doubling different voices of a triad at a distance of one octave is shown in Fig. 6.2 g. The best note to double in a major triad, according to calculated complex tonalness values, is the root (the first chord in the figure); either the root or the third may be doubled in the minor triad (the fourth and fifth chords). The lowest calculated (pure and complex) tonalness values result from doubling the third of the major triad (the second chord) and the fifth of the minor (the sixth chord). The calculated tonalness of the diminished triad is lowest when the conventional root is doubled. On the whole, these results are consistent with conventional doublings in Western harmony (Sect. 1.1.2).

Complex tonalness values in Fig. 6.2 tend to cover a wider relative range than pure tonalness values. This is to be expected, as pure tonalness depends only on the audibility of pure tone components, whereas complex tonalness additionally depends on the presence of harmonic pitch patterns between audible pure tone components. Theoretically, complex tonalness may be a more appropriate measure of musical consonance than pure tonalness, as almost all noticed (or perceptually relevant) tone sensations are of the complex kind.

6.1.5 Pitch Analyses

According to the model, a pure tone evokes weak complex tone sensations at subharmonic pitches (Sect. 2.4.5). The calculated pitch configuraton of a pure tone is graphed in Fig. 6.3 a. The calculated salience (probability of noticing) of the main (pure) tone sensation (C_4) is 0.78, that of the suboctave (complex) tone sensation (C_3), 0.13. The calculated multiplicity of the tone (1.3) applies only when the tone appears in isolation; in context, its multiplicity may be set equal to one, and calculated saliences may accordingly be divided by 1.3.

The complex tone (Fig. 6.3 b) has 16 harmonics whose auditory levels decrease with increasing harmonic number according to (4.8). According to the model, only the lower 10 harmonics are audible (Sect. 6.1.1), so only these influence the calculated salience of complex tone sensations; the irregular spacing of the harmonics 11 – 16 (caused by pitch categorization) has no effect on the calculation. The two highest calculated tone saliences are 0.55 (at the fundamental, C_4) and 0.26 (at the upper octave, C_5). The calculated salience of the upper fifth (G_4, a residue tone sensation) is only 0.06. So the calculated octave-ambiguity of the tone is much more pronounced than its calculated fifth-ambiguity. This is consistent with assumptions concerning octave equivalence and fifth relationships in Western music theory. As before, the calculated multiplicity of the tone (1.9) applies only when it appears in isolation; when the tone is perceived as a single whole, calculated saliences may be divided by 1.9.

According to the model, the main tone sensation of the pure tone in Fig. 6.3 a is more salient than that of the complex tone in part (b) of the figure, due to the relative absence of distracting subsidiary tone sensations in the pure tone. This is consistent with experimental results [Fastl and Stoll 1979]. However, the calculated *audibility* of the main tone sensation of the complex tone is much greater than that of the pure tone, as more sensory cues (harmonics) point to its existence. This distinction between salience and audibility resolves conceptual difficulties with the term *pitch weight* in Terhardt's model: pitch weight (as he defines it) is a measure of audibility (as defined here), not of salience (as defined here).

The octave-spaced tone (Fig. 6.3 c) has pure tone components at all the C's in the range of hearing, from C_1 (32 Hz) to C_9 (8.4 kHz). The tone sensations it evokes are mainly C's, although some weak (residue) tone sensations on F's, and some still weaker residue tone sensations on A*b*'s, are also output by the model. (This is because the third and sixth harmonics of an F, and the fifth and tenth harmonics of an A*b*, are all C's.) The experiments described in Sects. 5.3.3 and 5.6.3 provided tentative evidence for the existence of subfifth, but not subthird, tone sensations in octave-spaced tones; the latter may simply be too weak to be detected experimentally. The main tone sensation evoked by an octave-spaced tone, according to the model, has an equivalent frequency in the neighbourhood of 300 Hz, in agreement with Terhardt et al. [1986]. This absolute pitch effect is associated with spectral dominance (Sect. 6.1.1), and explains experimental results [Deutsch et al. 1984; Deutsch 1987] in which one of two octave-spaced tones a tritone apart is consistently heard to be higher or lower than the other. The calculated multiplicity of the tone is 2.9, i.e. considerably greater than that of an ordinary complex tone (Sect. 5.2.3). This suggests that octave-spaced tones are less likely than full complex tones to be heard as single entities, even in context.

According to the model, the most salient tone sensation of the minor triad $C_4 - Eb_4 - G_4$ corresponds to its lowest note, middle C (Fig. 6.3 d). Next in order of calculated salience are the tone sensations corresponding to the other two notes [Terhardt et al. 1982 a]. The root of the chord (C) corresponds to the most prominent chroma in the bass region. The chroma probability profile

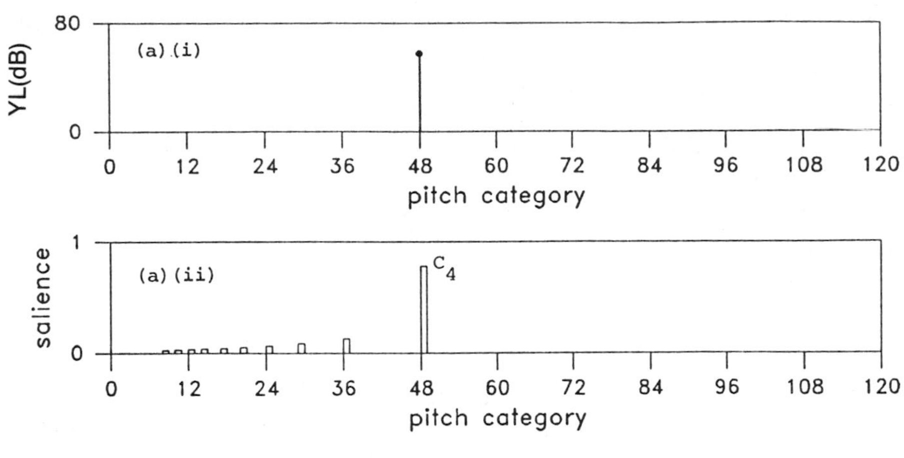

80
YL(dB)
0
(a) (i)
0 12 24 36 48 60 72 84 96 108 120
pitch category
1
salience
0
(a) (ii)
C4
0 12 24 36 48 60 72 84 96 108 120
pitch category

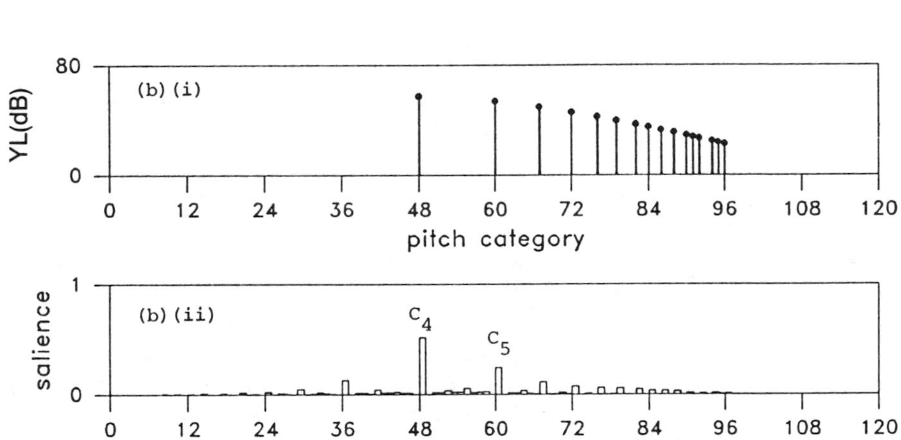

80
YL(dB)
0
(b) (i)
0 12 24 36 48 60 72 84 96 108 120
pitch category
1
salience
0
(b) (ii)
C4
C5
0 12 24 36 48 60 72 84 96 108 120
pitch category

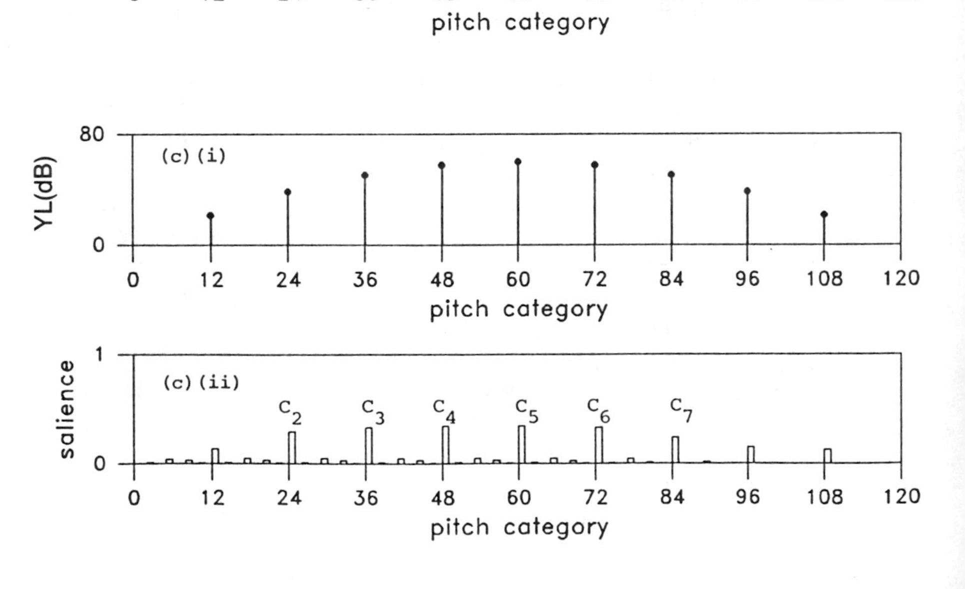

80
YL(dB)
0
(c) (i)
0 12 24 36 48 60 72 84 96 108 120
pitch category
1
salience
0
(c) (ii)
C2
C3
C4
C5
C6
C7
0 12 24 36 48 60 72 84 96 108 120
pitch category

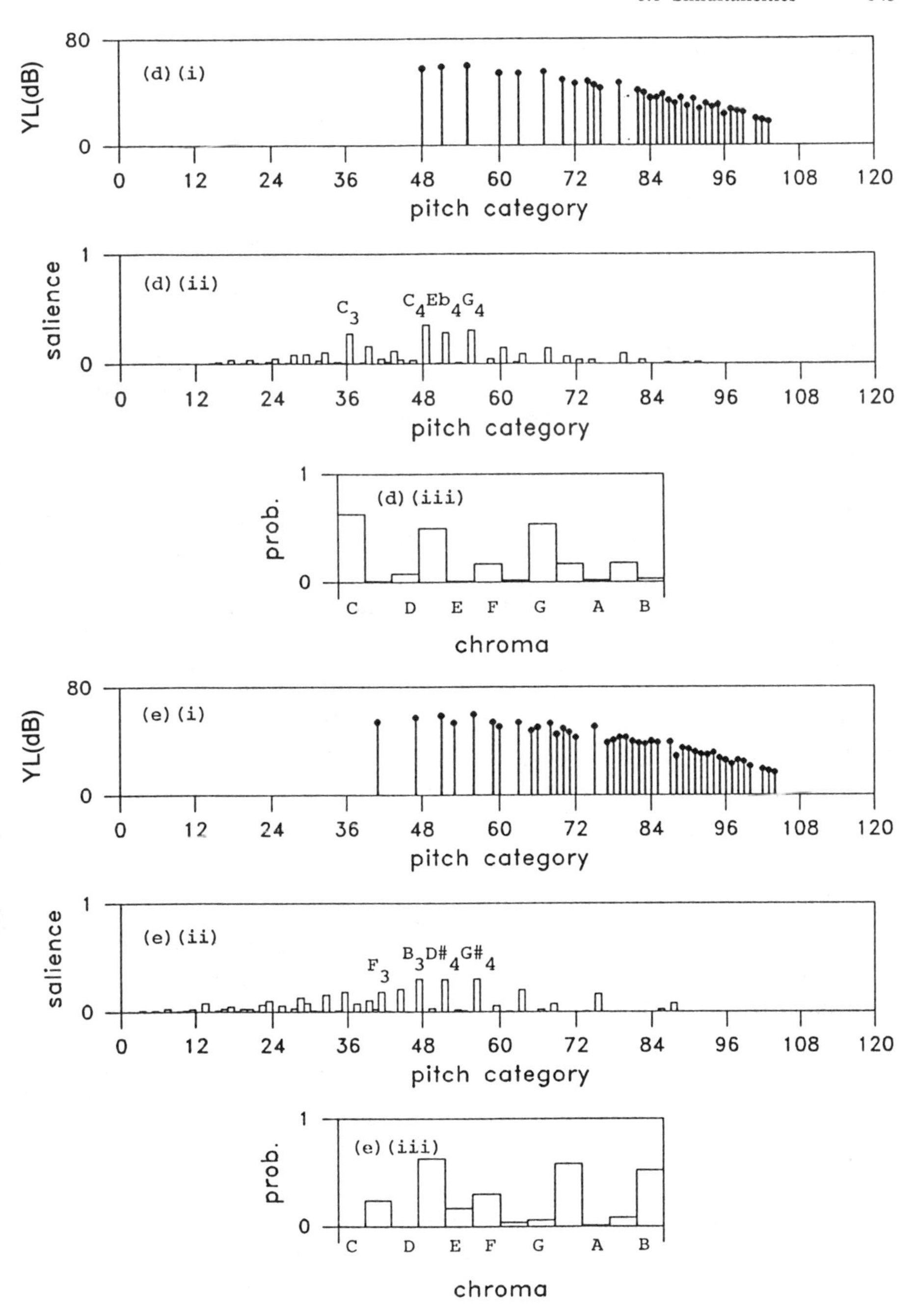

Fig. 6.3. Calculated pitch analyses of tones and chords. Auditory levels YL according to (4.6, 8, 10); saliences according to (4.26). (**a**) Pure tone on C_4. (**b**) Full complex tone on C_4. (**c**) Octave-spaced tone on C. (**d**) Minor triad $C_4-Eb_4-G_4$. (**e**) *Tristan* chord $F_3-B_3-D\#_4-G\#_4$. (*i*) Pure tone components. (*ii*) Tone sensations. (*iii*) Chroma probability profile

of the chord is slightly different from the chroma probability profile of its octave-generalized equivalent (Table 6.2b): it depends to a small extent on voicing (inversion, spacing, doubling).

The final sound simultaneity analyzed in this section is the musicologically infamous *Tristan* chord (Fig. 6.3e). Terhardt [1982] has suggested that the root of the Tristan chord is C#, as all its notes correspond to harmonics of C#. However, C# does not figure at all prominently in the present analysis. The three most salient tone sensations of the chord (corresponding to its upper three tones) are almost equally salient, suggesting that the root of the chord is quite ambiguous. The most salient tone sensations evoked by the chord in the bass region are E_2, $G\#_2$ and B_2 (pitch categories 28, 32 and 35). While these enhance the ease with which the Tristan chord resolves onto the E^7 chord which follows it (by enhancing the pitch commonality of the two chords), none is salient enough to represent the root of the chord: the most salient of the three, B_2, has a calculated salience of 0.18, which is relatively small in comparison to the salience (0.27) of the tone sensation evoked by the minor triad $C_4-Eb_4-G_4$ at C_3 (Fig. 6.3d), and not much bigger than the calculated salience (0.15) of the nearby tone sensation at $G\#_2$.

In this light, it is not surprising that music theorists disagree about the root and function of the *Tristan* chord. The chord is traditionally described as a French augmented sixth chord in A minor (F–B–D#–A) with a rising chromatic appoggiatura to the A (G#); it may also be interpreted as a chromatic alteration of the diminished seventh F–B–D–G# [Mitchell 1967; Knupp 1984]. In a psychoacoustical approach, it is a chord in its own right: a half-diminished seventh chord with no clearly defined root, set in an unusual context.

6.1.6 Chroma Salience and the Root

A *chord class* is specified by the intervals between the root and the other pitch classes of the other notes of the chord. For example, a major triad is specified by intervals of 4 and 7 semitones above the root. The root is normally understood in music theory to depend only (or primarily) on chord class: it is almost independent of voicing (inversion, spacing, doubling). By contrast, the pitch properties of different voicings of the same chord class (e.g. the saliences of particular tone sensations) can differ markedly [Terhardt et al. 1982a].

In order to investigate thoroughly the relationship between the root of a chord class and its calculated pitch properties, it would be necessary somehow to average the pitch properties of many different voicings of a particular chord class. A convenient and appropriate short-cut to this procedure is to build the chord out of *octave-spaced tones* (4.10).

Hindemith [1940] pointed out that *dyads* have roots, just as chords (triads, sevenths, etc.) do, and began his discussion of the roots of chords accordingly. Although there are twelve different dyads of ordinary complex tones within the

Table 6.2. Chroma probability profiles of simultaneities of octave-spaced tones. (Boldface values correspond to actual notes)

(a) Dyads

		C		D		E	F		G		A		B
m2	C D*b*	**0.80**	**0.77**	0.03	0.03	0.00	0.20	0.19	0.00	0.12	0.12	0.03	0.02
M2	C D	**0.79**	0.00	**0.81**	0.00	0.03	0.17	0.00	0.17	0.10	0.00	0.19	0.00
m3	C E*b*	**0.78**	0.02	0.03	**0.79**	0.00	0.34	0.00	0.00	0.50	0.00	0.02	0.12
M3	C E	**0.85**	0.00	0.08	0.00	**0.69**	0.15	0.02	0.00	0.09	0.15	0.02	0.00
P4	C F	**0.63**	0.09	0.02	0.02	0.00	**0.88**	0.00	0.02	0.08	0.00	0.23	0.00
T	C F#	**0.80**	0.00	0.27	0.00	0.13	0.20	**0.81**	0.00	0.26	0.00	0.02	0.21

(b) Triads

		C		D		E	F		G		A		B
maj.	C E G	**0.83**	0.00	0.05	0.05	**0.43**	0.15	0.01	**0.44**	0.05	0.15	0.01	0.00
min.	C E*b* G	**0.77**	0.01	0.02	**0.69**	0.00	0.27	0.00	**0.55**	0.28	0.02	0.01	0.06
aug.	C E G#	**0.73**	0.12	0.06	0.00	**0.74**	0.11	0.06	0.00	**0.74**	0.11	0.06	0.00
dim.	C E*b* G*b*	**0.68**	0.02	0.20	**0.64**	0.02	0.27	**0.69**	0.00	0.53	0.00	0.02	0.40
sus.	C F G	**0.76**	0.05	0.02	0.10	0.00	**0.77**	0.00	**0.51**	0.07	0.01	0.15	0.00

(c) Tetrads

		C		D		E	F		G		A		B
M7	C E G B*b*	**0.77**	0.00	0.04	0.18	**0.40**	0.10	0.09	**0.36**	0.07	0.13	**0.38**	0.00
m7	C E*b* G B*b*	**0.61**	0.01	0.01	**0.74**	0.00	0.17	0.03	**0.39**	0.25	0.01	**0.38**	0.04
M7	C E G B	**0.70**	0.01	0.04	0.06	**0.68**	0.09	0.02	**0.61**	0.03	0.23	0.01	**0.29**
∅7	C E*b* G*b* B*b*	**0.46**	0.01	0.11	**0.69**	0.01	0.14	**0.63**	0.00	0.41	0.00	**0.48**	0.25
d7	C E*b* G*b* B*bb*	**0.58**	0.02	0.46	**0.59**	0.02	0.46	**0.59**	0.02	0.46	**0.58**	0.02	0.46

(∅ denotes "half-diminished")

octave in Western music, there are only six different kinds of dyad made up of octave-spaced tones. A dyad of octave-spaced tones a fifth apart, for example, is simply a transposition of a dyad of octave-spaced tones a fourth apart. The six kinds of dyad correspond to the six *interval classes* (1, 2, 3, 4, 5 and 6 semitones) referred to by Forte [1964]. Calculated chroma probability profiles of the six different dyads of octave-spaced tones are presented in Table 6.2a.

According to the model, the only two of these for which one chroma is clearly more important than all the others are the dyads spanning the intervals of a major third (4 semitones) and a perfect fourth (5 semitones). Their main chroma correspond to their musical roots. In the case of the major second (2 semitones), the weak emphasis on the top note suggested by the model is in accord with an interpretation of the dyad as part of a dominant seventh chord. The tendency for the minor third (3 semitones) to be interpreted as part of a minor triad (i.e. for the lower note to be the root) is not reflected by the calculations. It is not clear whether this effect is due to cultural conditioning or to the role of harmonic overtones: the third harmonic of E*b* (omitted when the chord is built from octave-spaced tones) is octave-equivalent to the seventh harmonic of C. For the other intervals – the minor second (1 semitone) and the tritone (6 semitones) – the calculated saliences of the two chroma are about the same. Accordingly, these intervals are not assigned roots in music theory.

Calculated chroma probability profiles of some *triad* classes are shown in Table 6.2b. There are 20 possible triad classes [Forte 1964]; for reasons of space, only the five most common (or consonant) ones are analyzed here.

The calculated root of the C major triad, like its music-theoretical root, is clearly C. The supremacy of the root (C) of the C minor triad is jeopardized, but not overwhelmed, by the salience of the third (E*b*). The augmented, diminished and suspended triads are more ambiguous with respect to their roots. An additional candidate for the root of the C diminished triad (besides its actual tone chroma) is A*b* – a tone which, if physically present, would turn the triad into a major-minor seventh chord. This explains why the diminished triad on the leading note normally functions as a dominant seventh with missing root. The suspended triad ("sus") could be interpreted either as a suspended fourth chord (with root C) or a suspended second chord (with root F).

Cuddy and Badertscher [1987] asked children and adults with varying amounts of musical training and experience to rate how well octave-spaced tones on all twelve degrees of the chromatic scale went with broken (melodic) major and diminished triads. For the major triad, measured profiles peaked clearly at the conventional root; subsidiary peaks occurred at the other two notes, and at other notes in the associated major scale (with the exception of the leading note). For the diminished triad, profiles were essentially flat. These results agree qualitatively with the calculations presented in the table, suggesting that listeners familiar with Western music can extrapolate tonal implications of simultaneous tones (chords) to those of sequential tones (broken chords), see Sect. 3.4.3.

Schenker [1906] asserted that the diatonic scales derive from their tonic triads, and not vice versa. On the whole, the presented data are consistent with this view. The five most prominent chroma evoked by the C major triad (C–E–G) are C, E, F, G and A; the other notes of the C major scale, D and B, are not prominent. In the case of the C minor triad (C–E*b*–G), C, E*b*, F, G and A*b* are prominent; again, the second and seventh degrees of the scale, according to the model, are not particularly strong. On the other hand, scale degrees II and VII are the closest in pitch to the tonic. In short, scale degrees IV and VI are primarily *harmonically* related to the tonic *triad*, whereas VII and II are *melodically* related to the tonic *note*.

Chroma probability profiles of tetrads of octave-spaced tones are presented in Table 6.2c. Like the major triad, the major-minor seventh chord has an unambiguous root, according to both the model and music theory; this explains why it is so important in music theory, in spite of its relatively low tonalness (Sect. 6.1.4).

The calculated root of the minor seventh chord C–E*b*–G–B*b* is E*b*, suggesting that the chord is best regarded as an E*b* sixth chord. This conflicts with music theory and practice, in which C and E*b* are about equally likely to serve as functional roots of the chord. The minor seventh chord was just as common as the major sixth chord (e.g. IV^6 in a major key) during the Baroque, classical and romantic periods; both commonly preceded dominant harmony at perfect cadences. One might expect a slight bias in favour of the added sixth chord, in accordance with its more frequent appearance as a tonic chord (examples from Mahler and Debussy, as well as jazz and popular styles come to mind) but not as much as suggested by the calculations. The calculations could be improved by omitting the masking section of the model: when the chord is constructed from octave-spaced tones, C is masked more than E*b*, as it is closer to B*b* than E*b* is to C.

Like the minor seventh chord, the major seventh chord C–E–G–B has two possible roots: the conventional root (C) and the major third (E). However, E seldom acts as the root of this chord in music. Again, the masking algorithm is largely responsible for the model's underestimation of the salience of the conventional root (C).

The root of the half-diminished seventh chord C–E*b*–G*b*–B*b* is calculated to be E*b* rather than C. This agrees with music theory, in which IV^6 is preferred to II^7 in preparation for V^7 in a minor key (e.g. in Bach chorales); the half-diminished seventh chord is regarded as more stable in first inversion than root position, because the fifth above the bass is perfect rather than diminished. Note also that the minor sixth chord (e.g. E*b*–G*b*–B*b*–C) can function as a tonic (e.g. in Gershwin's *Summertime*), but (as far as I know) a half-diminished seventh cannot. Another possible root of the half-diminished seventh is the sub-major-third (A*b* in the example); this explains why the chord functions as a dominant (more precisely, as a dominant ninth with missing root) when it precedes the tonic triad in the key of D*b* major.

The diminished seventh is the most ambiguous chord of all, having potential roots not only at, but also just below, all its note chroma. The addition of a note a semitone (or minor ninth) below any of the notes of a diminished seventh turns it into a dominant minor ninth chord.

In summary, the root of a chord class is found to correspond well with the chroma with the highest calculated salience, when the chord is constructed from octave-spaced tones. This supports Terhardt's [1974a] claim that the root of a chord – although it is defined in cultural terms – has a strong sensory basis.

6.2 Progression

6.2.1 Pitch Commonality

The pitch commonality of a pair of simultaneities is the extent to which they evoke tone sensations falling in the same pitch categories (Sect. 3.2.3). It is quantitatively defined (Sect. 4.6.1) such that it equals 1 for two identical sounds.

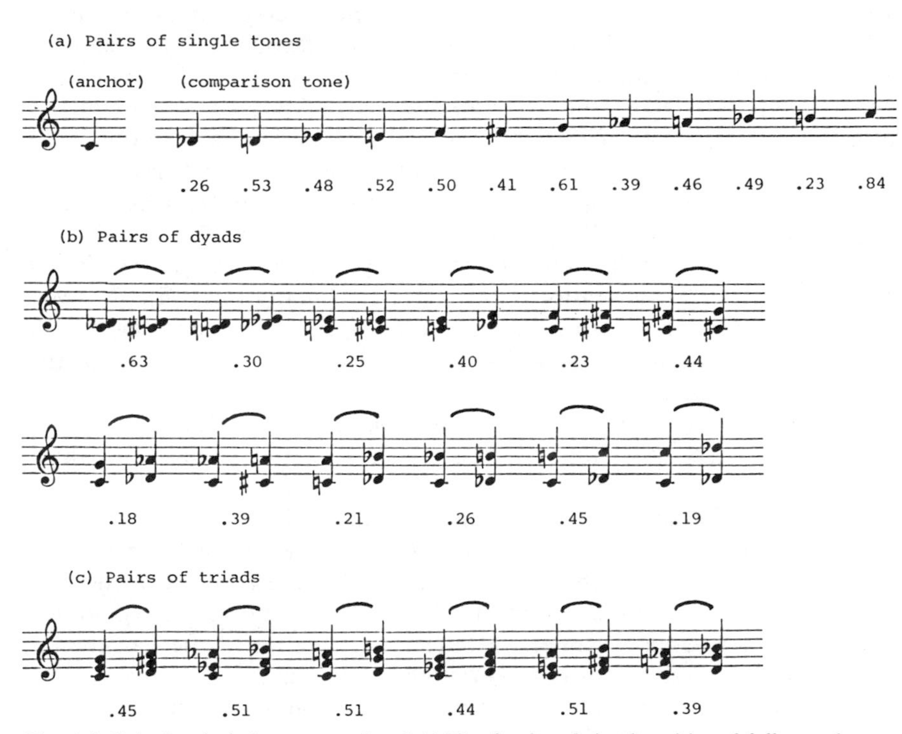

Fig. 6.4. Calculated pitch commonality C (4.29) of pairs of simultaneities of full complex tones (4.7, 8)

Figure 6.4a shows calculated pitch commonalities of sequential complex tones. The rank order of intervals obtained by this method, from lowest to highest pitch commonality, is M7, m2, m6, T, M6, m3, m7, P4, M3, M2, P5 and P8 (where M stands for major, m for minor, P for perfect and T for tritone). This order is similar to that obtained by ranking distances on the cycle of fifths: T, m2/M7, m3/M6, M3/m6, M2/m7, P4/P5, P8. The biggest differences between the two ranks involve the perfect fourth, which is stronger according to the cycle of fifths method, and the tritone, which is stronger according to the pitch commonality method.

In Fig. 6.4b, pairs of dyads in which voices move exactly in parallel over an interval of one semitone are compared for pitch commonality. The resultant rank order of intervals, from lowest to highest pitch commonality, is P8/P5, M6, P4, m3/m7, M2, m6/M3, T/M7, m2. This order resembles a rank from closest to farthest on the cycle of fifths. More interestingly, the intervals for which parallel motion through a semitone produces the lowest pitch commonality are the octave and fifth, because the salience of octave and fifth dyads is concentrated in relatively few pitch categories. This suggests a link between pitch commonality and the conventional avoidance of parallel fifths and octaves in mainstream Western harmony. Other factors associated with this convention include the fusion of tones in parallel fifths and octaves, which reduces their definition and contrapuntal independence, and a reaction against Medieval *organum* styles.

Figure 6.4c presents pitch commonalities of six pairs of triads moving in exact parallel motion through the interval of a major second. The pairs separate into two groups: those of relatively high pitch commonality (the second, third and fifth pairs) and those of relatively low pitch commonality (the others). Included in the second group are two combinations incorporating parallel fifths (the first and fourth).

A progression from one chord class to another may be *voiced* in different ways, i.e. in different inversions, spacings and doublings. In music theory, certain progressions are assumed to be strong regardless of how they are voiced, providing parts move through small steps, and parallels are avoided. Ideally, a psychoacoustical analysis of specific chord class relationships would involve averaging over a large number of acceptable voicings. This process may be short-cut by building chords from octave-spaced tones (as in Sect. 6.1.6).

Table 6.3a shows calculated pitch commonalities of pairs of major triads of octave-spaced tones at all possible chromatic intervals. One of the chords in each pair (the "anchor") is C major; the other is a major triad whose root is indicated in the top row of the table. The rank order of increasing pitch commonality is G*b*, D*b*, D, E*b*/E, F, C. The rank order of strength of harmonic relationships according to the cycle of fifths is G*b*, D*b*, E, E*b*, D, F, C. The relationship between C and D is weaker according to pitch commonality than it is according to the cycle of fifths, as triads a second apart have no notes in common. The major third relationship is stronger according to pitch commonality: major triads as major third apart have a common note (E in this

Table 6.3. Calculated pitch commonality of chords of octave-spaced tones

(a) Pairs of major triads (anchor: C major)

Comparison chord (major)	C	D*b*	D	E*b*	E	F	G*b*
Pitch commonality	1.00	0.26	0.39	0.51	0.52	0.63	0.21

Rank: G*b* D*b* D E*b* E F C

(b) Pairs of minor triads (anchor: C minor)

Comparison chord (minor)	c	c#	d	d#	e	f	f#
Pitch commonality	1.00	0.27	0.39	0.51	0.54	0.63	0.21

Rank: f# c# d d# e f c

(c) Mixed pairs of triads (anchor: C minor)

Comparison chord (major)	C	D*b*	D	E*b*	E	F	F#	G	A*b*	A	B*b*	B
Pitch commonality	0.78	0.35	0.27	0.81	0.30	0.60	0.26	0.59	0.75	0.24	0.45	0.45

Rank: A F# D E D*b*/B G F A*b* C E*b*

(d) Seventh chord and major triads (anchor: G^7, G – B – D – F)

Comparison chord (major)	C	D*b*	D	E*b*	E	F	F#	G	A*b*	A	B*b*	B
Pitch commonality	0.61	0.42	0.59	0.57	0.48	0.60	0.32	0.92	0.33	0.41	0.69	0.51

Rank: F# A*b* A D*b* E B E*b* D F C B*b* G

(e) Seventh chord and minor triads (anchor: G^7, G – B – D – F)

Comparison chord (minor)	c	c#	d	e*b*	e	f	f#	g	g#	a	b*b*	b
Pitch commonality	0.61	0.29	0.74	0.35	0.74	0.48	0.39	0.76	0.47	0.47	0.48	0.70

Rank: c# e*b* f# g#/a f/b*b* c d d/e g

case). The results for pairs of minor triads (part b of the table) are similar to those for pairs of major triads.

Three of the above comparison chords have no notes in common with the anchor chord: D♭, D and G♭. As these chords are constructed from octave-spaced tones, they do not have any frequencies in common, either. Yet, the D chord clearly stands in a stronger harmonic relationship to the C chord than D♭ or G♭ do. Bharucha and Stoeckig [1987] confirmed this experimentally, interpreting it as "evidence for activation spreading via a cognitive structure that links related chords" (p. 523). In a psychoacoustical approach, such harmonic relationships are initially explicable in purely sensory terms (pitch commonality), and are reinforced by cultural conditioning (Sect. 3.1.1).

Table 6.3c presents calculated pitch commonalities for pairs of triads in which one is major, the other minor. The strongest relationship is found between the C minor triad and its relative major (E♭). Next come the parallel major relationship (C), the submediant relationship (A♭), and the major dominant (G) and the major subdominant (F). Other relationships, according to the model, are much weaker. (Note that the G–c cadence is more fundamental than A♭–c in music theory, because only G–c clearly defines a key.)

Table 6.3d lists the pitch commonality of a G major-minor seventh chord (the dominant seventh of C) and all twelve major triad classes. Calculated pitch commonality is strongest for three common notes (G^7–G); these chords are so similar that they hardly invoke a feeling of progression. The second strongest pitch commonality is predicted for two common notes (G^7–B♭). This progression is quite common (e.g. II^7–IV or IV–II^7 leading to V^7 in the key of F major). The major triads F, D and C have about equal pitch commonality with G^7. This is consistent with musical practice: in the key of C, F and D often precede G^7 in perfect cadences, and C often follows G^7. The coherence of the progressions ♭VI–V^7 (A♭–G^7) and ♭II–V^7 (D♭–G^7) is not accounted for by these pitch commonality rankings. It may be explained instead by pitch commonality with other chords usually occurring in the same context (e.g. the pitch commonality of ♭VI with i and iv, and of ♭II with iv), and by pitch distance: the root of ♭VI is a semitone away from the tonally important scale degree V, and the root of ♭II is a semitone away from I.

Table 6.3e compares the chord G^7 with each of twelve minor triads. Apart from the parallel minor triad g, the most closely related minor triads to G^7 belong to the key of C major (d, e). Next comes b: G^7 functions as a German augmented sixth chord when it resolves to this chord, and is notated G–B–D–E#. Despite their chromatic relationship, G^7 and b have high pitch commonality, because they have two notes in common. The pitch commonality of G^7 and c (V^7 and *i* in C minor) is surprisingly high, considering that these chords have only one tone in common; the pitch commonality of G^7 with other minor triads with which it shares one common note (f, b♭ and g#) is much lower.

(a) Pairs of single tones

(b) Pairs of dyads

(c) Dominant-tonic progressions

Fig. 6.5. Calculated pitch distance D, (4.30), and D', (4.31), between pairs of simultane-ities of full complex tones (4.7, 8)

6.2.2 Pitch Distance

Two different formulations for the apparent overall pitch distance between sequential sounds are described in Sect. 4.6.2. The first, (4.30) for D, takes into account all tone sensations in each sound, and their calculated saliences. The second, (4.31) for D', takes into account only the actual complex tone components (notes) of a musical chord. Values of D and D' are compared for specific sound pairs in Fig. 6.5.

Figure 6.5a deals with pairs of single complex tones (notes) in musical intervals. Values of D' are simply equal to interval size in semitones. Values of D are smaller, due to overlapping of pitch configurations. For small intervals, D is less than half of the interval size in semitones. With increasing interval size, the magnitude of D accelerates relative to interval size, until the pitch distance between tones an octave apart exceeds half of the interval size.

Figure 6.5b shows calculated pitch distances between major third dyads. Here, the effect of overlapping pitch configurations is more pronounced: both formulations of pitch distance suddenly accelerate as the interval between the dyads exceeds a major third. These data are consistent with the conventional avoidance of this kind of part-crossing.

Figure 6.5c deals with three different voicings of a dominant seventh to tonic progression. Conventional theory recommends that the third of the dominant should rise by a semitone to the root of the tonic, and the seventh of the dominant should fall to the third of the tonic. Only the first of the three progressions conforms to both of these conventions; and only (4.31) for D' is consistent with music theory in this case. This confirms that D' is a better indicator of good voice leading (Sect. 4.6.2).

Values of D' in part (c) of the figure seem too large in comparison to values in parts (a) and (b). This could be rectified by reducing the effective salience of each tone component in (4.31) for chords of more than three notes, in line with the observation that it is difficult to attend to more than three tones at once (Sect. 6.1.3).

6.2.3 Tonicity

Section 3.4.3 dealt with the tonality of scales and melodies. The present section specifically investigates the tonality of *chord progressions*. The investigation is based on sensory properties of the chords according to the model in Chap. 4.

Musical experience suggests that the tonality of a Western chord progression is influenced by the following four factors. (i) The tonic pitch occurs more often than other pitches in a progression. (ii) The tonic pitch is the root of the chord formed by superposing the roots of chords in the progression (e.g. it is the root of the fifth between V and I: Sect. 3.4.3). (iii) The tonic chord stands in a strong harmonic relationship with other chords in the progression. (iv) The tonic chord is relatively consonant.

In my Ph.D. thesis [Parncutt 1987a], I attempted to determine the tonic of typical, octave-generalized chord progressions by means of systematic and (as far as possible) non-arbitrary procedures based upon each of the above criteria. None of the four criteria was found to predict the tonic with any generality when applied alone, for the following reasons.

Point (i) is not entirely true: the *dominant* is normally the note which occurs most often, as it is common to both tonic and dominant harmony. The tonic nevertheless sounds more important than the dominant, apparently because it is the root of the perfect fifth dyad formed by the tonic and dominant when they are heard simultaneously (Sect. 3.4.2). Point (ii) is inadequate by itself: the roots of suspended triads such as II–V–I and IV–V–I are ambiguous (Sect. 6.1.6), yet the corresponding progressions have relatively unambiguous tonics. Point (iii) produces similar ambiguities. Regarding (iv), consonance enhances the sense of finality associated with the tonic chord (see *time-span reduction preference rule 2* of Lerdahl and Jackendoff [1983]); but this criterion cannot by itself determine the tonic in progressions where all chords have about the same consonance (e.g. progressions of major and minor triads). In any case, the resolution of dissonance is not necessarily sufficient for the finality of a cadence [Matthews and Pierce 1980]. Evidence for this is the use of quite dissonant final (tonic) chords in jazz.

Predictions improved as more criteria were taken into account in the one procedure. It eventually became apparent that all four criteria needed to be ac-

Table 6.4. Calculated tonicity of chords in octave-generalized progressions

1st chord	Weight	2nd chord	Weight	3rd chord	Weight
C	0.67	G^7	0.54		
c	0.67	G^7	0.52		
C	0.66	F	0.64	G	0.60
c	0.64	f	0.62	G	0.54
C	0.64	G	0.63	d	0.62
c	0.62	d^o	0.58	G	0.57
C	0.63	G^7	0.53	D^7	0.52
c	0.63	Ab^7	0.52	G^7	0.49
F	0.63	G	0.62		
f	0.61	G	0.58		

Uppercase letters: chords with major thirds. Lowercase: minor. d^o denotes "d diminished": D – F – A*b*

counted for, if the tonic was to be predicted with any reliability, and the following procedure was developed. First, configurations of salience against pitch category were calculated for each chord in the sequence. Criterion (i) was accounted for by combining the configurations across time to yield the overall salience of each pitch in the chromatic scale ($P = 0, \ldots, 120$). Saliences were expressed as probabilities of noticing each pitch category at any time during the progression [after (4.28)]. Criterion (ii) was accounted for by applying the harmonic pattern-recognition procedure represented by (4.19) to the progression's overall configuration of salience against pitch. The output consisted of potential candidates for the "root of the progression", i.e. the root of both simultaneous and sequential pitches, as if they had all been heard simultaneously. Next, criterion (iii) was accounted for by calculating the pitch commonality of each chord in the progression with the array of candidates for the "root of the progression". Finally, criterion (iv) was accounted for by multiplying these pitch commonality values by tonalness values for the corresponding chords. The resultant value assigned to each chord was called its *tonicity.*

Results are shown in Table 6.4. The chord with the highest calculated tonicity corresponds to the tonic for all tested chord progressions whose tonics are clearly defined in music theory (but not for ambiguous progressions such as F–G and f–G). Moreover, progressions with relatively high maximum tonicity have relatively stable tonal structure.

The order of presentation of the chords of each example in the table is not specified. If the predictions of the described method were to be tested experimentally, they would need to be compared with results averaged over all possible orders of presentation of each progression. In order to account theoretically for order of presentation in specific cases, it would be necessary to consider such things as the decay of the "sensory trace" of each chord with time, and interference effects due to the presentation of successive chords (Sect. 2.2.3).

The described procedure takes both sensory and cultural aspects of Western tonality into consideration, suggesting that both sensory and cultural effects contribute to the perception of tonality and tonics in Western musics. Sensory effects include tone salience, tonalness and pitch commonality. Culture-specific aspects of the procedure include the assumption in point (iii) that listeners are familiar with the sounds of triads and the relationship between a triad and its root, and the particular way in which sensory effects are selected and combined in the procedure.

The above procedure is not completely reliable. For example, the tonics of the progressions c–f–G and C–G–d (rows 4 and 5 of the table) are relatively unambiguous (c and C respectively), but calculated tonicity values for these progressions do not clearly distinguish candidates for the tonic. The tonic of such progressions may have been disambiguated during the historical development of Western tonality, as the major and minor scales became familiar during the 16th and 17th centuries.

6.2.4 Implied Triad/Scale

A *key* may be regarded as a tonal system centred either on a chord class with a clear root (the tonic triad) or on a scale with a clear tonic (the tonic note) (Sect. 3.4.3). All twelve note chroma (pitch classes) may occur (as notes) in a given musical key [Goldman 1965]. What distinguishes one key from another is how often particular chroma occur. Chroma belonging to the prevailing scale occur more often than others (non-diatonic, or chromatic, chroma). Of the diatonic chroma, the most commonly occurring are the tonic and dominant.

Tonic triads and scales are not always played in their complete or typical forms: they need only be *implied* in order to produce a sense of key [Browne 1981]. Such implication may only be understood by a listener with extensive experience of music based on diatonic scales and triadic harmony. Several different triads and scales may be implied by a passage, so several different triads and scales may compete for the status of tonic.

Consider first the possibility that the tonic is that triad class which is most strongly implied by the progression. The pitch configurations of typical major and minor triads are familiar to the Western listener. Given octave equivalence, these pitch configurations may adequately be specified by *chroma salience profiles* (4.28), calculated directly from pitch analyses of major and minor triads of octave-spaced tones (Table 6.2). These profiles may be used as templates in a pattern recognition model (Sect. 2.3.3). If the tonic of a chord progression is an implied triad, the progression's chroma salience profile should correlate better with the chroma salience profile of the tonic triad than with that of any other major or minor triad. According to this hypothesis, however, the tonic of the progression CdG is A minor or E minor (left part of Table 6.5). These two triads go well with the triads in the progression, but only act as tonics in "modal" harmonic styles, without leading notes.

Table 6.5. Correlation of chroma probability profiles of octave-generalized progressions with profiles of triads and scales

Chords in progression			"Implied triad" method			"Implied scale" method		
C	G^7		C (0.77)	G (0.65)	e (0.60)	C (0.76)	F (0.64)	G (0.49)
c	G^7		c (0.84)	G (0.57)	C (0.51)	c (0.82)	E*b* (0.68)	B*b* (0.55)
C	F	G	C (0.65)	F (0.56)	G (0.51)	C (0.88)	F (0.75)	B*b* (0.54)
c	f	G	c (0.76)	f (0.54)	A*b* (0.47)	c (0.86)	E*b* (0.77)	A*b* (0.70)
C	d	G	a (0.61)	e (0.60)	C (0.58)	C (0.95)	F (0.82)	G (0.61)
c	d^o	G	g (0.47)	G (0.47)	c (0.30)	E*b* (0.86)	c (0.82)	A*b* (0.62)
C	D^7	G^7	C (0.62)	G (0.60)	D (0.43)	C (0.82)	G (0.73)	F (0.68)
c	Ab^7	G^7	c (0.80)	A*b* (0.55)	E*b* (0.43)	c (0.86)	E*b* (0.68)	A*b* (0.53)
F	G		F (0.61)	G (0.60)	d (0.60)	C (0.80)	F (0.64)	G (0.59)
f	G		f (0.61)	G (0.50)	F (0.37)	c (0.68)	E*b* (0.59)	f (0.59)

Uppercase: major chords and scales. Lowercase: minor. d^o denotes "d diminished": D – F – A*b*. Numbers in brackets are correlation coefficients

The role of leading notes in determining tonics may be accounted for by correlating the chroma salience profile of a progression with *scale* templates, in which each chroma is assigned a value of 1 (if it belongs to the scale) or 0 (if it does not). For example, the C major scale may be represented by the sequence 101011010101; C minor by 101101011001. (The last digit in each case corresponds to the leading note, B.) Results for representative chord progressions appear in the right part of Table 6.5.

On the whole, the implied scale method is successful. Occasional failures are due to the failure of the method to account for the different relative importances of different scale degrees. For example, the method assumes the leading note to be just as important as the tonic, and so predicts that the progression c – d^o – G (where the superscript o means "diminished") is in E*b* major instead of C minor. The leading note of both major and minor keys actually has relatively low harmonic importance, as reflected in music theory by its distance from the tonic along the cycle of fifths.

6.2.5 Key Profile

Krumhansl and Kessler [1982] measured the relative importances of the twelve chroma in a musical key, by comparing representative chords and chord progressions in a given key with single probe tones, presented just after each passage. The effect of pitch register was minimized by building both chords and probe tones from octave-spaced tones. Listeners, all of whom regularly played musical instruments, were "instructed to rate on a 7-point scale how well, in a musical sense, each probe tone fit into or went with the musical element just heard" (p. 342). Results (Fig. 6.6) were averaged over four different musical elements: an isolated tonic triad (I); the progression IV (subdomi-

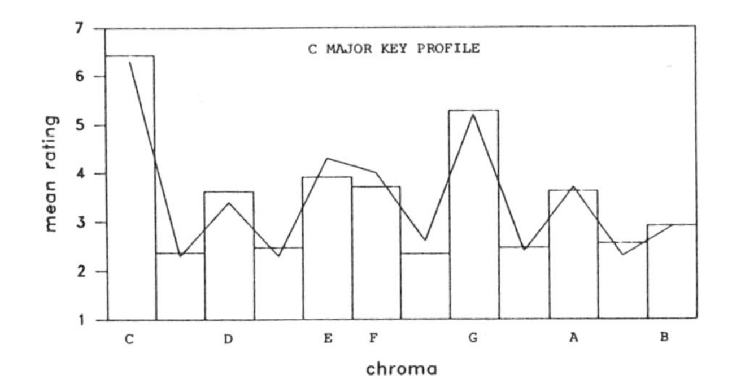

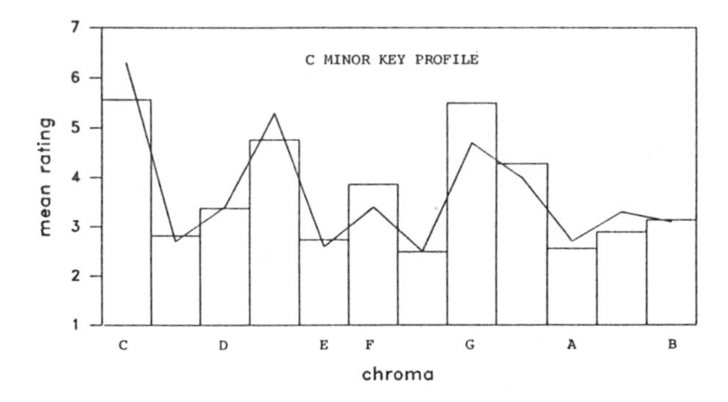

Fig. 6.6. Modelling of Krumhansl and Kessler's [1982] key profiles. *Zig-zag lines*: experimental results (see "Expt." rows in Table 6.6). *Histogram*: calculations (see "Calc." rows in the table) fit to experimental results by linear regression

nant), V (dominant), I; the progression II (supertonic), V, I; and the progression VI (submediant), V, I.

Krumhansl and Kessler supposed that "the hierarchy of tonal stability is acquired through experience with the structural relations that obtain in the music itself" (p. 364). They hypothesized two main factors which could lead to the acquisition of such a hierarchy. First, some chromatic scale degrees occur more often than others in diatonic music. These tend to be the same tones which are judged by listeners to "go well" with a musical passage. Second, the structure of tonal hierarchies bears some resemblance to the harmonic structure of single tones. This suggests that the acoustical properties of musical tones have influenced the evolution of harmony in Western music.

These explanations are not fully satisfactory. First, the most commonly occurring tone in a major or minor key is more often the dominant than the tonic (Sect. 6.2.3). Second, the relationship between the minor key hierarchy and the harmonic series is not clear (Sect. 1.2.3).

It turns out that the experimental results of Krumhansl and Kessler may be modelled quite closely by chroma tally profiles (Table 6.6). The effect of all the chords in the progressions on the experimental results may be taken into account by averaging over all four musical elements listed above, giving each element equal weight. This is achieved by adding up the calculated chroma salience profiles of six I's, three V's, one II, one IV and one VI triad. The

Table 6.6. Modelling of Krumhansl and Kessler's [1982] key profiles. (Boldface values correspond to actual notes)

(a) Calculation of major key template

		I					IV		V				
I	(maj.)	**1.57**	0.00	0.05	0.05	**0.55**	0.16	0.01	**0.54**	0.06	0.16	0.01	0.00
ii	(min.)	0.01	0.06	**1.30**	0.01	0.02	**1.06**	0.00	0.31	0.00	**0.73**	0.32	0.02
IV	(maj.)	**0.54**	0.06	0.16	0.01	0.00	**1.57**	0.00	0.05	0.05	**0.55**	0.16	0.01
V	(maj.)	0.16	0.01	**0.54**	0.06	0.16	0.01	0.00	**1.57**	0.00	0.05	0.05	**0.55**
vi	(min.)	**1.06**	0.00	0.31	0.00	**0.73**	0.32	0.02	0.01	0.06	**1.30**	0.01	0.02

Comparison of calculation (6I + ii + IV + 3V + vi) with experiment [Krumhansl and Kessler 1982]:

	I					IV		V				
Calc.	**11.5**	0.2	**3.7**	0.5	**4.5**	**3.9**	0.1	**8.3**	0.5	**3.7**	0.7	**1.7**
Expt.	**6.4**	2.2	**3.5**	2.3	**4.4**	**4.1**	2.5	**5.2**	2.4	**3.7**	2.3	**2.9**

Correlation coefficient: $r = 0.986$

(b) Calculation of minor key template

		I					IV		V				
i	(min.)	**1.30**	0.01	0.02	**1.06**	0.00	0.31	0.00	**0.73**	0.32	0.02	0.01	0.06
ii	(dim.)	0.02	0.48	**1.07**	0.02	0.22	**0.96**	0.02	0.31	**1.03**	0.00	0.71	0.00
iv	(min.)	**0.73**	0.32	0.02	0.01	0.06	**1.30**	0.01	0.02	**1.06**	0.00	0.31	0.00
V	(maj.)	0.16	0.01	**0.54**	0.06	0.16	0.01	0.00	**1.57**	0.00	0.05	0.05	**0.55**
VI	(maj.)	**0.55**	0.16	0.01	**0.54**	0.06	0.16	0.01	0.00	**1.57**	0.00	0.05	0.05

Comparison of calculation (6i + ii + iv + 3V + IV) with experiment [Krumhansl and Kessler 1982]:

	I					IV		V				
Calc.	**9.6**	1.1	**2.8**	**7.1**	0.8	**4.3**	0.0	**9.4**	**5.6**	0.3	1.3	**2.1**
Expt.	**6.3**	2.7	**3.5**	**5.4**	2.6	**3.5**	2.5	**4.8**	**4.0**	2.7	3.3	**3.2**

Correlation coefficient: $r = 0.941$

results correlate well with Krumhansl and Kessler's measured tone profiles (kindly supplied by Dr. Krumhansl).

The histogram in Fig. 6.6 is obtained when the calculated key profiles are mapped by linear regression onto experimental profiles. Assuming that the calculated tone profiles well reflect the sensory properties of major and minor keys, discrepancies between calculations and experimental results reflect the influence of cultural conditioning. This is expected to have been considerable due to the musical orientation of both the listeners and the experimental task in the experiments of Krumhansl and Kessler. Bigger discrepancies occur for the minor key, supporting the music-theoretical idea than the minor is somehow less "natural" (i.e. less sensorially based) that the major, or that it represents a "distorted" version of the major (Sect. 3.4.2). The supremacy of the tonic over the dominant, which has a clear sensory basis in the case of the major key, appears to be largely cultural in the minor case: the sensory basis for the root of the minor triad is relatively weak (Table 6.2b). The mean experimental response for the third degree of the minor scale considerably exceeds the calculated response, perhaps because the minor third of the minor key is the most important chroma distinguishing it from the major key. The

Table 6.7. Correlation of chroma probability profiles of octave-generalized progressions with key profiles

Chords			Correlations		
C	G^7		C (0.93)	G (0.70)	c (0.62)
c	G^7		c (0.93)	C (0.63)	E♭ (0.50)
C	F	G	C (0.89)	F (0.76)	G (0.62)
c	f	G	c (0.92)	f (0.57)	A♭ (0.53)
C	d	G	C (0.83)	G (0.75)	F (0.67)
c	d^o	G	c (0.81)	g (0.53)	E♭ (0.51)
C	D^7	G^7	C (0.85)	G (0.81)	F (0.52)
c	$A♭^7$	G^7	c (0.92)	A♭ (0.56)	E♭ (0.47)
F	G		F (0.73)	C (0.67)	G (0.64)
f	G		c (0.64)	f (0.60)	C (0.50)

Uppercase: major chords and keys. Lowercase: minor. d^o denotes "d diminished": D – F – A♭

low salience of the leading note of both major and minor keys is consistent with its distance from the tonic on the cycle of fifths, and with rules preventing its doubling in harmony.

A key profile may be regarded as a cross between a triad template and a scale template: key profiles account not only for the importance of the tonic triad, but also for the importance of the prevailing scale, in the determination of musical keys. Krumhansl and Kessler determined the key of specific passages of music by correlating their experimentally determined key profiles with profiles of the passages (obtained by the same experimental method). They also used their key profiles to produce a multidimensional map of relationships between musical keys (generalizing the music-theoretical idea that scales are perceived to be related if they have tones in common), and traced the tonal movement of specific passages through the space.

Predicted tonics of typical chord progressions according to the method of implied keys, using only calculated versions of Krumhansl and Kessler's tone profiles, are shown in Table 6.7. The results are correct in all cases where the tonic is clearly defined in music theory. This is not surprising, as the key templates were themselves calculated from chroma profiles of similar progressions.

The method of implied key is restricted to chord progressions, as the term "key" itself presupposes triadically based harmony. The method does not always correctly determine the key of unaccompanied melodies. The key of an unaccompanied melody in Western music may instead be found by the method of implied scales (Sect. 6.2.4). Alternatively, the key of a melody may be regarded as the key of its implied accompaniment, i.e. the key of an implied chord progression.

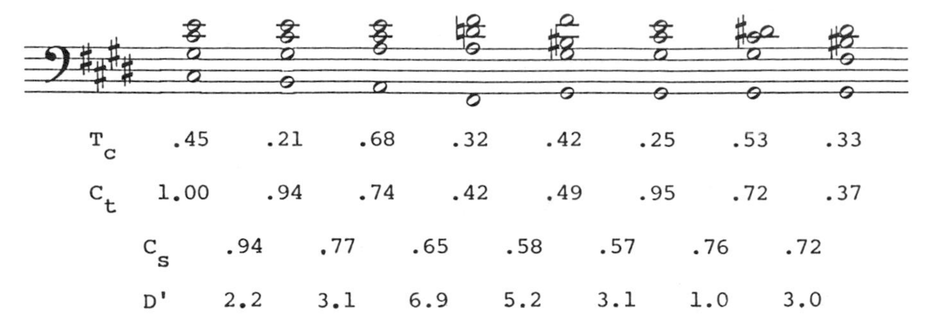

Fig. 6.7. Analysis of the opening eight bars of Beethoven's Sonata Op. 27 No. 2. T_c: Complex tonalness (4.22) of each chord. C_t: Pitch commonality (4.29) of each chord with the first chord (the tonic). C_s: Pitch commonality of successive chords. D': Pitch distance (4.31) between successive chords

6.3 Pieces

6.3.1 Analysis

The model in Chap. 4 may be applied to the analysis of specific chord progressions, including underlying chord progressions of polyphonic passages.

An appropriate harmonic reduction of the first four bars of Beethoven's *Moonlight Sonata* is notated in Fig. 6.7. Complex tonalness (a measure of the consonance of a chord, see Sect. 6.1.4) varies considerably during the progression. It is lowest for the second chord. The unexpectedly low apparent dissonance of this chord may be explained either psychoacoustically, in terms of the relatively high pitch commonality and low pitch distance between the chord and its neighbours (Sect. 3.2.1), or music-theoretically, in terms of voice leading (the B in the bass being merely a passing note). Complex tonalness is highest for the third chord, being the only root position major triad in the progression.

Pitch commonality between each chord and the tonic chord (defined here to be the opening chord) is lowest for chords whose notes do not all belong to the standard heptatonic scale defined by the key signature: the fourth chord in the progression (a Neapolitan sixth) and the fifth and last chords (different voicings of the dominant seventh, with sharpened thirds). In all these cases, low pitch commonality is associated with dissonance, tension, and expectation of resolution.

Pitch commonality between successive chords is consistently high ($C > 0.5$), reflecting the general harmonic cohesion of the passage. Pitch distances between successive chords increase to a maximum toward the middle of the phrase, as the progression moves away from the tonal centre, and then fall back to low values in preparation for the return of the tonic (after the last chord in the example).

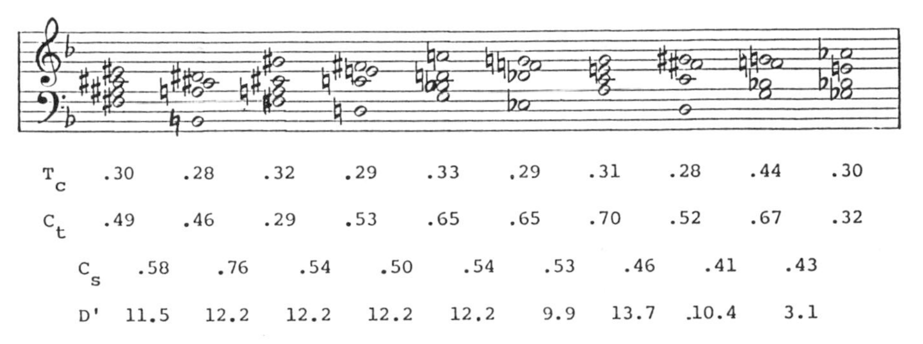

Fig. 6.8. Analysis of an arrangement of a progression from *The Girl From Ipanema* by Jobim, de Moraes and Gimbel. T_c: Complex tonalness (4.22) of each chord. C_t: Pitch commonality (4.29) of each chord with $F_3-A_3-C_4-F_4$ (the tonic). C_s: Pitch commonality of successive chords. D': Pitch distance (4.31) between successive chords

According to the progression's chroma probability profile (Table 6.8), the chroma most likely to be noticed in the progression are the tonic and dominant, C# and G#. Next in line are the minor third (E) and perfect fourth (F#); the minor sixth (A) is also well represented. The key profile of C# minor, in which C#, G#, E, A and F# rate highly, therefore correlates well with the chroma probability profile of the progression, in spite of the roughly equal saliences of I and V (C# and G#), see Fig. 6.6. The high correlation coefficient ($r > 0.8$) reflects the clarity of the tonal structure of the passage; apparently, the Neapolitan sixth chord (chord 4) does not disrupt the prevailing C# minor tonality. The key of E major is predicted to be the second most strongly implied key of the passage; the piece modulates (in passing) to E major a few bars after the end of this phrase.

The excerpt analyzed in Fig. 6.8 is an arrangement (by the author) of a progression from *The Girl From Ipanema* by Jobim, De Moraes and Gimbel. The analysis shows that the techniques developed in this study can be applied to quite complex harmonies and could be useful in jazz research.

Complex tonalness (and hence consonance) is generally lower, and varies considerably less, in this example than in the Beethoven. Instead, musical unity seems to be maintained by *homogeneity* of texture and consonance, suggesting a kind of trade-off between homogeneity and consonance. Complex tonalness is highest for the second-last chord, the G minor seventh: this is the only chord with octave doubling. The progression could perhaps be made more homogeneous by reducing the consonance of this chord, e.g. by shifting the upper voice of the chord up a tone.

The tonic of the sequence is defined for the purpose of analysis to be the chord $F_3-A_3-C_4-F_4$, a chord which nicely precedes and follows the sequence. Pitch commonality with the tonic chord is generally lower than in the Beethoven example. Again, the last chord has low pitch commonality with the tonic, increasing expectation that it will resolve to the tonic (which it in fact

Table 6.8. Octave-generalized analysis of progression in Fig. 6.7

Chroma probability profile of entire progression:

C		D		E	F		G		A		B
0.73	0.99	0.70	0.74	0.95	0.26	0.95	0.23	1.00	0.91	0.45	0.75

Correlation of chroma probability profile with diatonic templates:

	C		D		E	F		G		A		B
major	−0.39	+0.32	+0.14	−0.42	+0.59	−0.62	+0.39	−0.39	+0.11	+0.56	−0.74	+0.43
minor	−0.45	+0.81	−0.37	−0.05	+0.02	−0.25	+0.52	−0.57	+0.39	+0.13	−0.29	+0.10

Rank: c# E A f# B F#/g# C# D a G# b e d# f a# d C/G E*b* c g F

Table 6.9. Octave-generalized analysis of progression in Fig. 6.8

Chroma probability profile of entire progression:

C		D		E	F		G		A		B
0.94	0.98	0.95	0.89	0.93	0.93	0.97	0.97	0.93	0.92	0.95	0.72

Correlation of chroma probability profile with diatonic templates:

	C		D		E	F		G		A		B
major	+0.04	+0.37	+0.06	−0.07	−0.41	+0.06	+0.34	+0.04	−0.06	−0.07	+0.14	−0.44
minor	+0.05	+0.20	+0.29	−0.23	−0.51	+0.31	+0.13	+0.28	−0.45	−0.04	+0.28	−0.31

Rank: D*b* f b*b*/g c#/d B*b* D/F c G/C A*b* E*b*/A e*b* b B E g# e

does). The fifth (G minor-add-9), sixth (Eb^9, seventh (Am7) and ninth (Gm7) chords have the highest pitch commonality with the tonic: all these commonly occur in jazz in the key of F major. The third chord (F# minor-add-9) is quite remote from the key of F major, and also has the lowest pitch commonality with the tonic.

Pitch commonality between successive chords is lower than in the Beethoven example, but never falls below 0.4, despite large distances traversed on the cycle of fifths. The lowest value occurs between the third-last and second-last chords. The relationship between these chords is maintained in other ways than by pitches in common: the pitch distance between them is relatively small (no voice moves through more than a semitone), and the rising fourth in the bass is familiar to musical listeners. Pitch distance between successive chords is higher, but more consistent, than in the Beethoven example; again, pitch distance is lower just before the return to the tonic.

The consistently high chroma probabilities in the profile (Table 6.9) indicate that all twelve chroma are normally noticed at some time during the progression. The chroma least often noticed is B, lying diametrically opposite to F (the song's tonic) on the cycle of fifths. The correlation between the profile of the calculated key of this progression (D*b* major) and the chroma probability profile of the passage is low ($r<0.4$), implying a weak tonal structure (the passage is clearly not in D*b*!). F minor ranks highly in the list of calculated keys, suggesting that contact with the tonic key of the piece (F major) is not completely lost during this tonal adventure. A more interesting and relevant picture of the changing tonality of the passage (including passing modulations to F# major/minor and G minor) could be obtained by looking at a few chords at a time, or by allowing the effect of previously heard chords to decay exponentially with a time constant comparable to the duration of auditory sensory memory (Sect. 2.2.3).

6.3.2 Composition

The twentieth century is unique in the history of Western music. It is the only century in which the most commonly heard art music was composed in previous centuries. Despite many decades of exposure of Western audiences to a range of non-diatonic styles, old-fashioned diatonic tonality remains the most popular way of organizing musical pitch (Sect. 1.1.4).

Diatonic tonality is in many ways an arbitrary, culture-specific system, but (as this study has shown) it also has a strong sensory basis. This basis appears to be responsible for the survival of diatonic tonality over four centuries, in spite of considerable and repeated changes and developments in Western musical styles. It is also responsible, at least in part, for the introduction of Western tonal harmony to non-Western cultures [Nettl 1986].

The current study extends the work of Terhardt and his colleagues in isolating and quantifying sensory aspects of music perception. The point has been reached where it should be possible to apply this knowledge in consistent ways in order to create new tonal styles – styles whose sensory basis is similar to that of existing tonalities, but whose arbitrary, 'cultural' aspects are different.

Consider the composition of chords for a particular group of instruments or voices. The first limitation on such chords is the pitch range of each instrument or voice. Another limitation (built into Western notation, and into the model in Chap. 4) is that pitches are confined to semitone categories.

These two limitations, taken together, allow for a very large number of different chords. This number may be reduced to a manageable level by limiting the sensory consonance (roughness, tonalness) of chords (or other simultaneities), resulting in a kind of *vocabulary* from which homorhythmic progressions may be composed. The size of the vocabulary could be further reduced either by further manipulation of psychoacoustical parameters, or by the composer's personal choice. Relatively consonant vocabularies of this kind would resemble existing diatonic chord vocabularies; vocabularies of relatively low consonance would be less familiar, and so more likely to have interesting applications in composition.

A possible way of composing a progression from a given vocabulary of simultaneities might be to select the first simultaneity at random, and then to select subsequent simultaneities so that the pitch commonality and pitch distance between sequential pairs fell within prescribed limits. In order to give progressions more unity, it may be useful to prescribe a kind of tonal centre, defined as an array of salience against pitch category (Sect. 6.2.5), and to specify limits on pitch commonality between individual simultaneities and the tonal centre.

These techniques are similar to mathematical (including serial) techniques advocated by some twentieth century composers (e.g. Xenakis, Boulez), in that they make the composer's job easier: part of the compositional process becomes more or less automatic. The techniques described here have the additional advantage that they are based not on arbitrary or intuitive principles but on established psychoacoustical theory and data.

To be successful, new styles should grow out of existing, accepted styles, and should be introduced gradually to their audiences. New music needs to be introduced in appropriate contexts; for example, atonal styles may succeed in film soundtracks but fail in the concert hall. New music needs to be performed on appropriate instruments; for example, atonal electronic styles may succeed where atonal orchestral styles fail. These considerations hold regardless of the extent to which new styles may be based on universal aspects of auditory perception.

The techniques described above are not supposed to have any particular aesthetic value. As pointed out by Lerdahl and Jackendoff [1983], "Both tonality and contemporary techniques have produced masterpieces; both have produced trash" (p. 301). The primary source of the quality of a piece of music has always been (and, hopefully, will always remain) the personal contribution of the composer.

Glossary

The entries below are intended primarily to clarify the meaning of terms used in the main text. They are not supposed to be rigid definitions, nor are they necessarily valid or applicable outside the context of the present study. Many of the entries are based on, but not necessarily the same as, definitions used in acoustics [American Standards Association 1960; Beranek 1972], psychoacoustics [Zwicker 1982; Moore 1982], psychology [Drever 1964] and musicology [Apel 1970; Scholes 1970]. Some terms have been coined or redefined to clarify explanation and discussion of new and "revisited" material in the present study: these are marked with an asterisk (*).

Absolute pitch. Pitch of a pure tone (component) measured by equivalent frequency (e.g. "same pitch as a 60 dB, 300 Hz pure tone") or by pitch category (e.g. "D_4"); ability to identify absolute pitch to within a specified range of error (see perfect pitch).

Algorithm. Systematic procedure for the evaluation of a mathematical function; quantitative model.

Ambient pressure. Air pressure in the absence of sound.

Amplitude. Peak value of a function, e.g. peak sound pressure of a pure tone.

Amplitude spectrum. Amplitudes of the spectral components of a function, as a function of their frequencies.

Analysis. Spectral analysis, analytical listening, or musical analysis.

Analytical listening. Mode of listening in which: pure tone components are well discriminated; pure tone components are heard in preference to complex tone components; several tone sensations are heard simultaneously; and/or pitch relationships between sounds are influenced by pitch commonality in preference to pitch proximity.

Anchor sound. Invariant sound (compared with a number of different sounds).

Apparent. Perceived.

Arithmetic mean. Of two values: half their sum. Of *n* values: one *n*th of their sum.

Atonal. Kind of music in which harmonic relationships are avoided (e.g. 12-tone music) so that there is little or no sense of tonal centre.

Attitude. Mode of listening, e.g. holistic versus analytic, or spontaneous versus voluntary.

Attribute. Property, e.g. the pitch, loudness or timbre of a sound sensation.

* **Audibility.** Audibility of a pure or complex tone (component); depends on audible level (cf. Terhardt: *pitch weight*).

* **Audible level.** Level of a pure tone (component) above masked threshold (Terhardt: *SPL excess*); depends on auditory level and masking.

* **Auditory level.** Level of a pure tone (component) above the threshold of audibility in the absence of masking.

Auditory nerve. Nerve picking up information from the vibration of the basilar membrane in the cochlea.

Auditory stream. Stream.

Auditory system. Psychophysiological system concerned with hearing; the ear-brain system.

Augmented. Of a diatonic interval: one semitone bigger than the corresponding major or perfect interval. Opposite of diminished.

Augmented fourth (A4). Interval of 6 semitones; tritone; enharmonically equivalent to a diminished fifth.

Augmented fifth (A5). Interval of 8 semitones; enharmonically equivalent to a minor sixth.

Augmented triad. Triad with a M3 and an A5 above the root.

Augmented sixth. Chord with a M3 and an A6 (=m7) above the root.

Baroque. Period of Western music history extending from about 1600 to about 1750 (Monteverdi to Bach).

Bark. Unit of critical-band rate equal to one critical bandwidth.

Basilar membrane. Membrane in the cochlea (inner ear) enabling auditory spectral analysis.

Bass. Lowest note of a chord; low frequency or pitch register.

Bell-like sound. Sound whose spectrum is is like that of a bell (i.e. like a harmonic series, but with added, missing and/or out of tune components).

Beat frequency. Modulation frequency.

Beats, beating. Amplitude modulation of the waveform of a sound, usually due to the superposition of the waveforms of pure tone components of similar frequency and similar amplitude.

Carrier frequency. The mean frequency of two beating pure tone components.

Category. Subdivision of a continuum (e.g. real numbers in the range 4.5–5.5 fall in the category "5, to the nearest whole number"); group of related words (e.g. oak, wattle and gum); range of sensory attributes (e.g. shades of blue).

Categorical perception. Perception in terms of labelled categories. For example, Western musicians hear intervals in the range 2.5–3.5 semitones as minor thirds.

Cf. Compare.

Chord. Complex musical sound comprising three or more simultaneous notes, (complex tones); triad, tetrad (e.g. seventh chord), etc.

Chord class. Chord whose notes are specified only by their chroma (pitch classes), e.g. C–E–G instead of $C_4 - E_4 - G_4$ or $G_3 - E_4 - C_5$. Could be in any inversion, spacing or doubling (cf. voiced chord).

Chord progression. Sequence of chords; homorhythmic progression of sounds made up of complex tones (notes).

Chroma. Pitch class, e.g. "C" instead of "C_4" (middle C), "C_7", etc.; pitch without specification of octave register; interval in semitones between a pitch category and the nearest C below (e.g. the chroma of C_3 is 0, of Db_5 is 1, of F_4 is 5, etc.).

Chroma cycle (or circle). Music-theoretical construct of chroma resembling a clock-face, with C at twelve o'clock, C# at 1, D at 2, etc.

* **Chroma salience.** Measure of the perceptual importance of a particular chroma in a musical sound or sequence, as perceived by an average or "ideal" listener.

* **Chroma tally.** Calculated average number of times a chroma is noticed; measure of chroma salience.

* **Chroma probability.** Calculated probability that a chroma will be noticed; measure of chroma salience.

Chromatic scale. Scale with twelve roughly equally spaced pitch categories per octave; scale in which each pitch category has a well-tuned upper and lower octave and fifth.

Classical. Period of Western music history extending from about 1750 to about 1800 (Haydn, Mozart, early Beethoven, etc.).

Close position. Voicing of a chord class in which the lowest and highest notes are less than an octave apart.

Cochlea. Part of the inner ear; spiral-shaped cavity in the petrous bone housing the basilar membrane.

Combination tone. Tone produced by the simultaneous of other tones; pure tone component produced by non-linear interaction (distortion) among other pure tone components in an acoustical transducer (e.g. a loudspeaker, or the ear).

Complex sound. Sound whose pressure waveform is not sinusoidal, and whose spectrum therefore contains more than one pure tone component (e.g. a musical chord). (N.B.: This is essentially the same as the American Standards Association's definition of a complex *tone*).

* **Complex tonalness.** Measure of tonalness; the audibility of the most audible complex tone sensation of a sound.

Complex tone. Sound whose pressure waveform is (approximately or exactly) periodic (but not sinusoidal, see pure tone).

Complex tone component. Complex tone which is part of a sound (e.g. a note in a chord).

* **Complex tone sensation.** Single tone sensation normally accompanying the perception of a complex tone or complex tone component (see virtual pitch).

Components. Waveforms (usually periodic) which add to produce another waveform (usually, of a real sound).

Conditioning. Exposure to regular or invariant aspects of the environment, and its effect on perception. The term is often understood to refer to more specific processes by which "unnatural" or "abnormal" responses are elicited by particular stimuli.

Confidence interval. Mathematical interval within which the mean of a set of numbers is expected to lie with a specified probability (e.g. 95% confidence interval).

Consonance. How well the tones of a simultaneity, or the sounds in a sequence, "sound together", depending on roughness, tonalness, pitch commonality, pitch distance, context, familiarity and cultural conditioning.

Context. Here, all sensations which are simultaneously available to be perceived in a particular "chunk" of sensory memory.

Contour. The "ups and downs" in pitch of a melody, as opposed to the sizes of the intervals between the tones.

Contrapuntal. Polyphonic.

Correlate. Establish the degree to which two sets of numbers are related (implying that they represent the same or similar measurements or properties).

Correlation coefficient. Measure of the relationship between two sets of numbers, ranging from 1 (if one set of numbers is an exact, positive linear transformation of the other) through 0 (the numbers are totally unrelated) to -1 (one set is an exact, negative linear transformation of the other).

Critical band. Maximum range of frequencies over which the ear is like a single band-pass acoustic filter (so loudness is independent of bandwidth); at wider ranges, it is like a bank of band-pass filters (so loudness increases with increasing bandwidth).

Critical bandwidth. Width of a critical band (in semitones or Hz), equal to about 3 semitones above 500 Hz, and 50–100 Hz below 500 Hz. Contains a constant number of pitch difference thresholds.

Critical-band rate. Pure tone height.

Cultural. Due to (or specific to) a particular society or culture (as opposed to universal).

Cycle (or circle) **of fifths.** Music-theoretical construct of chroma in which neighbours span perfect fifth and fourth intervals. In semitones: 0, 7, 2, 9, 4, 11, 6, 1, 8, 3, 10, 5 and back to 0. Note names: C, G, D, A, etc.

dB. Decibel, unit of level.

Degree. Step (of a scale); pitch category.

Diatonic scale. Heptatonic scale comprising 5 intervals of a tone (whole steps) and 2 of a semitone (half steps) in each octave, such that half steps are not adjacent. Examples: standard heptatonic scale, mode, major scale, minor scale.

Difference threshold. Just noticeable difference, difference limen; smallest perceptible physical change in a stimulus.

Diminished. Of an interval: one semitone smaller than the corresponding minor or perfect interval. Opposite of augmented.

Diminished fifth (d5). Interval of 6 semitones; tritone; enharmonically equivalent to an augmented fourth.
Diminished triad. Triad with a m3 and a d5 above the "root".
Diminished seventh. Interval enharmonically equivalent to a major sixth (9 semitones); chord class with a m3, d5 and d7 above the "root".
Dissonance. Opposite of consonance.
Distortion. Undesired modification of a (physical) sound signal due to limitations of a transducer (e.g. a microphone or loudspeaker), filter, or amplifier.
Dominant. Diatonic scale step V, a perfect fifth above the tonic.
Dominant seventh. Major-minor seventh chord whose root is on the dominant.
Doubling. The simultaneous voicing of a chroma in different octave registers.
Dyad. Simultaneity of two complex tones (notes).

Enharmonic (or enharmonic equivalent). Having different musical note names but corresponding to the same chromatic scale step (e.g. C sharp, D flat).
Equal temperament. Tuning or intonation in which semitones cover equal frequency level intervals (one twelfth or an octave).
Equivalent frequency. Measure of pitch; frequency of a standard reference tone whose pitch is the same as that of a particular tone sensation.
Envelope (or amplitude envelope). Shape of a graph of amplitude against time.
Evoke. (Of a stimulus): produce (a sensation).

Fast Fourier Transform (FFT). Specific, efficient spectral analysis algorithm.
Fifth. Perfect, diminished or augmented fifth (usually, perfect).
Filter. Device which modifies an input signal such that, on spectral analysis, it contains more energy in certain spectral frequency ranges and less in others.
Flat, flatten. Lower in pitch, usually by a semitone.
Fourier analysis. Spectral analysis.
Fourth. Perfect, augmented or diminished fourth (usually perfect).
Frequency. Of a periodic waveform: number of cycles per unit time (e.g. per second); reciprocal of period.
Frequency level. Logarithmic measure of frequency, in semitones (or octaves) above 16.35 Hz (so that A_4, 440 Hz, has a frequency level of exactly $4 \times 12 + 9 = 57$ semitones).
Free parameter. Adjustable variable in an (otherwise fixed) algorithm.
Full complex tone. Complex tone with all pure tone components (harmonics), perhaps up to some specified limit of harmonic number or frequency.
Function. Mathematical relationship or operation with a single output value (or array of values) for each possible input value (or array of values).
Functional harmony. System of harmonic analysis in which chord roots are specified by their diatonic scale steps (I, II, III, etc.) and chords are interpreted as functionally equivalent to either I, IV or V.
Fundamental. First harmonic; lowest pure tone component of a full complex tone.
Fundamental frequency. Waveform frequency.

Gamelan. Indonesian gong orchestra.
Geometric mean. Of two numbers: the square root of their product.

Half-diminished seventh. Chord class with intervals m3, d5 and m7 (3, 6 and 10 semitones) above the (conventional) root.
Half step. Semitone.
Harmonic. Whole multiple of a specified number. *n*th harmonic: pure tone component whose frequency is (close to) *n* times the waveform (fundamental) frequency of a complex tone.
Harmonic complex tone. Complex tone whose pure tone components are exactly harmonic; complex tone which is exactly periodic (over some finite duration).
Harmonic meaning. Of a tonal sound: meaning associated with the harmonic relationship between the sound and its context.
Harmonic minor scale. Scale whose pitch categories lie 2, 3, 5, 7, 8 and 11 semitones (M2, m3, P4, P5, m6, M7) above the tonic.
Harmonic relationship. Perceived relationship between musical sounds involving intervals (e.g. P8, P5, M3) between specific pitch categories (cf. melodic relationship).
Harmonic series. Series of numbers of the form a, 2a, 3a, ...
Harmonicity. How closely the spectral frequencies of a sound correspond to a harmonic series, or how closely the pitches of the pure tone sensations of a sound correspond to those of a typical complex tone.
Harmony. General term embracing consonance (especially of simultaneities) and conventions for writing tonal progressions.
Hear out. Hear, by analytic listening, something which is not normally noticed (e.g. a harmonic of a complex tone).
Heptatonic scale. Scale with seven pitch categories per octave.
Hertz (Hz). Unit of frequency; cycles per second.
Histogram. Bar chart; kind of graph.
Holistic. Opposite of analytical. Normal mode of perceiving.
Homorhythmic progression. Chord progression whose voices all move in the same rhythm (i.e. without passing notes; note against note). Here, a sequence of sounds of similar loudness and timbre (each of which lasts long enough to be clearly and separately perceived) whose overall duration is short enough that the progression may be held in auditory sensory memory (cf. polyphonic progression).
Hz. Hertz.

Integer. Whole number.
Intensity. Of a sound: amount of energy transmitted per unit time, per unit area perpendicular to the direction of propagation.
Interval. In mathematics: segment of a one-dimensional scale lying between two specified values (see confidence interval). In music: distance between two pitch categories (in semitones); inclusive number of scale steps in a

diatonic scale between two notes (e.g. major third, augmented fourth, perfect fifth).

Intonation. Frequency adjustment of a note (within its pitch category) in music performance (see just, Pythagorean, equally tempered).

Inversion. Voicing of a chord specified by the chroma of the bass note – root position, first inversion (third in the bass), second inversion (fifth in the bass), etc. – or interval which, when added to another interval, makes an octave (e.g. a fifth is the inversion of a fourth).

Impressionism. Western school of musical composition (late 19th, early 20th century) whose best-known exponent was Debussy.

Just intonation. Intonation (or tuning) in which all intervals are expressed as combinations of octaves (2:1), fifths (3:2), and major thirds (5:4), so that the frequencies of tone pairs stand in simple, small-whole-number ratios with factors 2, 3 and 5. E.g. a just minor third has a frequency ratio of $(3/2)/(5/4) = 6/5 = 3 \times 2/5$.

Key. Musical pitch framework characterized by certain chord progressions (of which the dominant-tonic progression is the most important) and a specific hierarchy of pitch classes (headed by the tonic and dominant).

kHz. Kilohertz; thousands of cycles per second.

Leading note. Note a semitone away from (usually, below) a root or tonic; diatonic scale step VII.

Level. Logarithm of the ratio of a quantity to a reference quantity of the same kind; threshold level, auditory level, masking level, audible level or (usually) sound pressure level.

Loudness. Attribute of auditory sensation by which different sensations may be ordered on a scale extending from "soft" to "loud".

Main pitch. Pitch of the main tone sensation of a sound.

Main tone sensation. Most audible, salient or prominent tone sensation in a sound.

Major scale. Scale whose pitch categories lie 2, 4, 5, 7, 9 and 11 semitones (M2, M3, P4, P5, M6, M7) above the tonic in each octave.

Major second (M2). Interval of 2 semitones.

Major third (M3). Interval of 4 semitones.

Major sixth (M6). Interval of 9 semitones; inversion of a minor third; chord class with a M3, P5 and M6 above the root.

Major seventh (M7). Interval of 11 semitones; inversion of a minor second; chord class with a M3, P5 and M7 above the root.

Major-minor seventh (Mm7). Chord class with a M3, P5 and m7 above the root; dominant seventh.

Masked threshold. Threshold of audibility in the presence of maskers.

Masker. Sound (e.g. a pure tone component) which masks other sounds (e.g. other pure tone components).

Masking. Complete or partial "drowning out" of one sound by another.

Masking level. Degree to which a pure tone component is masked by other components; difference between the auditory level and audible level of a pure tone component.

Mean. Of two values: arithmetic (or geometric) mean. Of a continuous function: area under the function over a specified interval, divided by the length of the interval.

Mediant. Diatonic scale step III, a third above the tonic.

Melody. Sequence of tones in music perceived as related or as a whole due to their limited range of pitch, loudness and timbre (cf. stream).

Melodic minor scale. Minor scale with a major sixth and seventh when ascending, and a minor sixth and seventh when descending.

Melodic relationship. Perceived relationship between two sounds in music by virtue only of their pitch proximity (cf. harmonic relationship).

Melodic stream. Stream.

Microtonal. Relating to division of the octave into more than twelve pitch categories, or into categories less than a semitone wide.

Minor scale. Harmonic or melodic minor scale; heptatonic scale including pitch categories 2, 3, 5 and 7 semitones (M2, m3, P4, P5) above the tonic.

Minor second (m2). Smallest musical interval; one semitone.

Minor third (m3). Interval of 3 semitones.

Minor sixth (m6). Interval of 8 semitones; inversion of a major third; chord class with a m3, P5 and M6 above the root.

Minor seventh (m7). Interval of 10 semitones; inversion of a major second; chord class with a M3, P5 and m7 above the root.

Mode (or scale mode, or church mode). Version of the standard heptatonic scale centred around (and named after) one note (the final). For example, "white notes" on D constitute the *Dorian* mode.

Modern. Period of Western music history extending from about 1900 to the present.

Modulation. In acoustics: periodic change in the amplitude or frequency of a sound (beating). In music: change of key during a piece.

Modulation frequency. Frequency of the (smoothed) envelope of a waveform. The difference between the frequencies of two beating pure tone components.

Multidimensional scaling. Algorithm for summarizing data on the (apparent) similarity, relatedness or differentness of pairs within a set (e.g. a set of musical sounds). The output of the algorithm specifies the positions of the elements of the set in an *n*-dimensional space (where *n* may be specified by the user), such that distances between the elements in the space correspond roughly to their (apparent) differentnesses.

Music theory. Theory of musical structure (including composition and analysis) based primarily on the notes of musical scores.

Musical analysis. Analysis of the musical/tonal functions of tones, chords, key areas, etc., in a passage of music.

Musician. Musical performer.

Neurophysiological. Concerning brain function.
Noise. Sound whose spectrum contains continuous distributions (bands) of energy (rather than pure tone components); undesirable, disturbing or annoying sound.
Non-harmonic sound, non-harmonic tone. Complex sound which is not a tone, i.e. whose pure tone components do not conform to (or approach) a harmonic series of frequencies.
Notate. Encode sound in the form of instructions to a musical performer.
Notation. Musical performance instructions.
Note. Instruction to play a tone at an approximately specified time and frequency. (The actual onset time and frequency of a tone in performance depend on the categorical perception of pitch and time, and on context: roughness, rubato, etc.).
Notice. Consciously perceive.

Octave. Distance between two tones or frequencies corresponding to a frequency ratio of 2:1; frequency level difference of 12.0 semitones; musical interval covering eight diatonic scale steps (inclusive); distance between sequential tones perceived to lie an octave apart (see octave stretch).
Octave equivalence (or octave generalization). Assignment of the same names to notes an octave apart; assumption that notes an octave apart have the same harmonic function.
Octave stretch. Effect by which octaves in music correspond to frequency ratios slightly larger than 2:1.
Octave-generalized chord. Chord class.
Octave-spaced tone (or Shepard tone). Complex tone whose pure tone components are separated by octaves and cover most or all of the pitch range.
Optimal fit. Optimal closeness of two sets of data, minimizing the root-mean-square difference between corresponding data points.
Organum (or parallel organum). Medieval plainsong style in which parts move by parallel fifths or fourths.
Overtone. Harmonic other than the fundamental; pure tone component other than the lowest.

Pa. Pascal; newtons per square metre.
Parallel octaves, fifths. These occur when two voices a fifth or fourth apart move in parallel through the same chromatic interval.
Peak amplitude. Maximum displacement of a waveform from its mean or zero value.
Pentatonic scale. Scale with five pitch categories per octave.
Perceive. Detect, recognize or identify something (compare "notice"), usually while interacting with the environment.
Perfect fifth (P5). Interval between the first and fifth degrees of a major or minor scale, or between the second and third harmonics of a complex tone; 7 semitones.

Perfect fourth (P 4). Interval of 5 semitones; inversion of the perfect fifth.

Perfect octave (P 8). Octave.

Perfect pitch. Ability to identify the absolute chromatic pitch category of a musical tone; absolute pitch to the nearest semitone.

Perfect unison (P 1). Zero semitones.

Period. Time interval between repetitions of a periodic waveform.

Periodic function. Function that repeats itself at regular (time) intervals.

Periodic sound. Sound whose waveform is periodic.

Phase. Position of the point at which a periodic waveform passes through zero (relative to some specified point).

Phase spectrum. Phases of spectral (pure tone) components, expressed as a function of their frequencies.

Pitch. Attribute of a tone sensation by which it may be ordered on a scale from low to high.

Pitch ambiguity. Property of a sound having various possible pitches (when perceived holistically).

Pitch category. Range of pitch (about a semitone wide) within which the harmonic meaning of a musical note or tone stays the same.

Pitch class. Chroma.

* **Pitch commonality.** Degree to which two sounds are perceived to have pitches in common, depending on the number and saliences of tone sensations falling in common pitch categories.

Pitch configuration. Pattern of tone salience as a function of pitch.

Pitch distance. Overall apparent distance in pitch between two sounds (see melodic relationship).

Pitch prominence. Essentially the same as audibility, salience (of a pure tone).

Pitch proximity. Overall apparent closeness in pitch of two sounds; opposite of pitch distance.

Pitch relationship. Extent to which two sounds of similar loudness and timbre are perceived to be similar or related by virtue of their pitch configurations. Depends on harmonic relationship (pitch commonality) and voice leading (pitch proximity).

Pitch shift. Slight change in the pitch of a tone due to change in level or the presence of masker(s).

Pitch weight. Essentially the same as (tone) audibility.

Polyphonic progression. Contrapuntal progression; music consisting of two of more simultaneous lines or melodies.

Preference. The extent to which an observer "likes" (or evaluates as "good") a stimulus presented e.g. in an experiment.

Present. Play a sound in an experimental context.

Pressure. Force per unit area; instantaneous pressure relative to ambient pressure.

Pressure waveform. Waveform of sound pressure against time; the most basic physical description of a sound.

Probe tone. Tone to which a listener refers in order to answer a question about some other sound in an experimental trial.

Profile (tone profile, key profile). Graph (or histogram) of salience against chroma.

Program. Computer (software) implementation of an algorithm.

Progression. Sequence of sounds which are perceptually relatable to each other and may be stored in sensory memory.

Psychoacoustics. Psychophysics of hearing.

Psychophysics. Study of relationships between (physical) stimuli and the sensations (experiences) they evoke.

Pure intonation. Just intonation.

* **Pure tonalness.** Measure of tonalness dependent on the number and audibility of the pure tone components of a sound.

Pure tone. Tone whose pressure waveform is sinusoidal.

Pure tone component. Spectral component; partial.

* **Pure tone height.** Proportional sensory measure of the height (pitch) of a pure tone or pure tone component, measured in critical bandwidths above the lower threshold of pitch.

* **Pure tone sensation.** Tone sensation evoked solely by a pure tone or pure tone component (see spectral pitch).

Pythagorean intonation (or tuning). Hypothetical intonation (or tuning) in which all intervals are expressed as combinations of pure fifths (3:2) and octaves (2:1), so that frequencies of pairs of tones stand in simple, whole number ratios in which only factors of 2 and 3 appear. E.g. a Pythagorean major third has a frequency ratio of $(9/8)\times(9/8) = (81/64) = 3^4/2^6$.

Rating. Judgment or estimate by a participant in an experiment.

Register. Numerical label for the octave in which a tone, note or pitch falls, ranging from 0 to 9. The register from middle C (262 Hz) to the B above (494 Hz) is called register 4.

Relative pitch. How high a tone sounds relative to another, measured either on a continuous scale (pure tone height) or on a categorial scale (using musical intervals). Also, the ability to recognize musical intervals.

Renaissance. Period of Western music history extending from about 1450 to about 1600.

Residue pitch. Pitch of a residue tone sensation; pitch corresponding to a missing fundamental.

Residue tone. Complex tone with most lower harmonics present but with no fundamental.

* **Residue tone sensation.** Complex tone sensation whose pitch does not correspond to that of a pure tone component.

rms. Root-mean-square.

Romantic. Western music of the nineteenth century (late Beethoven, Chopin, Brahms, Liszt, Wagner, Mahler, . . .).

Root. "Generating" note or chroma upon which thirds are superposed in order to construct a chord in music theory; chroma relative to which the other chroma of a chord are perceived.

Root position. Inversion of a chord in which the bass note is the root.

Root-mean-square (rms). Average value of a function (e.g. a waveform) calculated by taking the square root of the mean of the square of the function.

Roughness. Sensation associated with beating at frequencies in the range 20–300 Hz.

Rubato. Controlled fluctuation of musical tempo.

Salience. Perceptual importance or prominence. Of a stimulus or environmental object: probability of being noticed. Of a sensation: probability of occurring (i.e. being experienced).

Sample. Particular (output) value of a function (such as a waveform), often taken at regular (input) intervals.

Scale (categorical or musical). Set of pitch categories in a progression or a piece of music which are used more often than others, or which act as a perceptual frame of reference. Scales normally include between 5 and 12 pitch categories per octave, and repeat (in the same pitch relationships) in different octaves.

Scale (sensory or psychophysical). Continuum for the measurement of a psychophysical parameter (such as pitch) in which equally sized intervals correspond to equal numbers of difference thresholds.

Scale step (or scale degree). Pitch category belonging to a scale; degrees of the diatonic scales are labelled I, II, . . . , VII.

Semianechoic. Not completely anechoic, i.e. absorbing (rather than reflecting) almost all sound so that echoes and reverberation are very weak.

Semitone. Unit of frequency level; twelfth part of an octave.

Sensation. Experience accompanying perception.

Sensory. Universal and perceptual; due to universal aspects of the physiology of hearing and the human auditory environment.

* **Sensory memory.** Spontaneous memory, in the absence of awareness or noticing (at the actual time of the "memorized" event).

* **Tonicity.** Of a chord in an octave-generalized progression: a measure of the probability that the chord will be perceived as the tonic, calculated on the basis of sensory properties of the progression.

Seventh. Chord with a third, fifth and seventh above the root; interval between the first and seventh notes of a diatonic scale. Major, minor or diminished.

Sharpen. Raise in pitch, usually by a semitone.

Short-term memory. Sensory (spontaneous) memory.

Significant. In statistics: most unlikely to have occurred by chance (e.g. $p<0.05$ means that the probability of a given result occurring by chance is less than 5%).

Simultaneity. Sound; simultaneous group of tones (complex tone, dyad, chord, bell sound, etc.), usually of short duration.

Sinusoidal. Mathematical function of the form $y = a \sin(bx+c)$, where a, b and c are constants.

Sixth. Major, minor or augmented sixth.

Smoothness. Lack of roughness.

Sound. Simultaneity.

Sound pressure. Oscillatory part of pressure; overall pressure minus ambient pressure.

Sound pressure level (SPL). Level of (rms) sound pressure relative to a reference sound pressure of 0.00002 Pa; an increase in SPL of 20 dB corresponds to a multiplication of pressure by a factor of 10.

Spacing. Aspect of voicing involving the sizes of the pitch intervals between the notes of a musical chord.

Spectral analysis. Mathematical procedure by which a function (e.g. a waveform) is expressed as a sum of individual sinusoidal functions over a specified analysis interval ("window"). The output of a spectral analysis consists of an amplitude and a phase spectrum.

Spectral component. Sinusoidal function output by spectral analysis; pure tone component.

Spectral dominance. Effect by which the pure tone components with the greatest influence on the pitch properties of complex sounds lie in the (approximate) frequency range 250–2000 Hz – the range which is most important for the perception of vowels in speech.

Spectral pitch. Pitch of a pure tone sensation.

Spectrum. Any function of frequency (e.g. amplitude spectrum, phase spectrum).

SPL. Sound pressure level.

Spontaneous. Without conscious effort or intent; not voluntary.

Standard deviation. Measure of the spread of a set of numbers, determined by a standard mathematical formula; distance away from the mean of a set of numbers (e.g. numbers intended as measurements of the same thing) within which about two-thirds of the numbers lie.

* **Standard heptatonic scale.** Scale corresponding to the white notes of the piano, or transpositions thereof; scale of seven pitch categories per octave corresponding to seven neighbouring chroma on the cycle of fifths. (Differs from *major scale* in that no tonic is specified).

* **Standard pentatonic scale.** Scale corresponding to the black notes of the piano, or transpositions thereof; scale of five pitch categories per octave corresponding to five neighbouring chroma on the cycle of fifths.

Standard reference tone. Pure tone of frequency 1 kHz and variable SPL used for loudness measurements, or pure tone of 60 dB (SPL) and variable frequency used for pitch measurements.

Stimulus. Physical entity (e.g. environmental state or process, or source thereof) which may be perceived, especially in a psychophysical experiment.

Stream. Sequence of tones spontaneously perceived as related or as a whole due to their limited range of pitch, loudness and timbre, or perceived to stem from a single source (due to auditory familiarity with that source); auditory stream; melodic stream; sensory basis for melody.

Style (or genre). Arbitrary, culture-specific way of organizing music.
Subdominant. Diatonic scale degree IV, a fourth above the tonic.
Subharmonic. Whole submultiple of a particular number (e.g. 2.5 is the 4th subharmonic of 10).
Subharmonic pitch. Pitch whose equivalent frequency is approximately subharmonic to some other frequency (such as that of a pure tone); e.g. the subharmonics of C_5 are C_4, F_3, C_3, $A\flat_2$, F_2, D_2, C_2, ...
Submediant. Diatonic scale degree VI, a sixth above (or a third below) the tonic.
Subthreshold. Below the threshold of audibility; inaudible.
Successive. Neighbouring and sequential.
Supertonic. Diatonic scale degree II, a second above the tonic.
Synthesize. Produce music electroacoustically.
Synthetic. Electroacoustically produced.

Tetrad. Chord of four different complex tones (notes), usually of different chroma (e.g. seventh chord, sixth chord).
Threshold. Value of a physical stimulus parameter (e.g. SPL, frequency) where the experience of a sound categorically changes (e.g. the sound becomes audible, the sound has pitch, . . .).
Threshold level. SPL at threshold of audibility.
Threshold of audibility. Threshold sound pressure (defined for an average "ideal" listener) below which a pure tone is inaudible, expressed as a function of its frequency.
Threshold of pitch. Lowest (20 Hz, E_0) or highest (16 kHz, C_{10}) audible pitch.
Timbre. Tone colour; attribute of a tone sensation by which different tone sensations of the same pitch and loudness may be differentiated.
Tonal. Of a sound in psychoacoustics: evoking pitch (i.e. tone sensation/s). Of a passage of music: exhibiting tonality.
Tonal relationship. Harmonic or melodic relationship, or both.
Tonality. Pitch structure in music in which some pitches (in particular, the tonic) are more important (salient, stable) than others; that aspect of musical structure involving pitch relationships.
Tonalness. The extent to which a sound evokes pitch, i.e. audible tone sensations; pure or complex tonalness; sonority, sonorousness.
Tone. Sound which evokes a tone sensation; approximately or exactly periodic sound in the audible range of frequencies; sound whose various possible pitches mostly belong to a single pitch class or chroma (e.g. full complex tone, octave-spaced tone, residue tone).
Tone component. Part of a sound which, if played by itself, would be a tone.
Tone salience. Salience of a tone or tone sensation.
Tone sensation. Auditory sensation having (one, unambiguous) pitch; other attributes include loudness or salience, timbre, and apparent duration.
Tonic. Most important sound, tone, pitch or chroma of a progression, acting as a perceptual reference relative to which other sounds, tones, notes, pitches or chromas are perceived; diatonic scale step I.

Transducer. Physical system converting sound into/from some other form of energy/information (e.g. microphone, loudspeaker, ear).
Transpose. Shift all notes/tones through the same (chromatic) musical interval.
Triad. Chord of three different complex tones (notes), usually of different chroma (e.g. major triad, minor triad).
Trial. Part of a psychophysical experiment, during or after which an observer makes a single rating or set of ratings and, as a result, a single piece or set of data are recorded.
Tritone. 6 semitones; 3 whole tones; augmented fourth or diminished fifth.
Tuning. Adjustment of (fundamental) frequency in music.
Twelve-tone music. Atonal chromatic music in which all twelve chroma (pitch classes) appear about equally often; based on "tone rows" (particular orders of the twelve chroma).

Unison. Interval of zero semitones.
Universal. Applying to most or all of the human race.

Virtual pitch. Pitch of a complex tone sensation.
Voice leading. Voicing, with emphasis on melodic relationships (pitch distances) between sequential sounds.
Voiced chord. Chord whose notes are assigned to specific octave registers (cf. chord class).
Voicing. Realization of an octave-generalized chord or chord sequence; allocation of notes of specified chroma to particular registers; inversion, spacing, and doubling.

Waveform. Function whose mean value over a specified range is zero but whose value at particular points (and hence its rms value) is not zero. Of a tone: graph of sound pressure against time.
Whole step. 2 semitones.

Glossary of Symbols

A	Audibility of a (pure or complex) tone component
A_c	Audibility of a complex tone component
A_{max}	Maximum tone audibility over all P in a simultaneity
A_p	Audibility of a pure tone component
AL	Audible level in dB of a pure tone component; level in dB above masked threshold (cf. YL below)
c	Chroma (pitch class) (number of semitones above C)
C	Pitch commonality of two simultaneities
cb	Critical bandwidth (or Bark)
D	Apparent pitch distance between two simultaneities
dB	Decibels; (logarithmic) unit of SPL, TL, YL, ML, AL
e	Base of natural logarithms ($\simeq 2.72$)
f	Frequency in Hz
FL	Frequency level (in semitones above 16.35 Hz)
H_p	Pure tone height (critical-band rate) in cb
Hz	Hertz; cycles per second
i	Arbitrary integer variable
int	Integer part (e.g. int $\{2.6\} = 2$)
I	Harmonic interval (in semitones above the fundamental)
k	Free parameter in model, as follows:
k_M	Masking parameter: gradient of the masking pattern of a pure tone in dB/cb (the higher k_M, the more components are audible)
k_R	Pitch relationship perception parameter: measure of how much pitch commonality influences sequential pitch relationships, as opposed to pitch proximity
k_S	Simultaneity perception parameter: measure of how analytically a simultaneity is heard (the higher k_S, the higher M)
k_T	Tone perception parameter: measure of how analytically individual tones are heard (the higher k_T, the more pure tone components are heard in preference to complex tone components)
$\log_{10}$	Logarithm to base 10
M	Multiplicity; number of simultaneously noticed tones
M'	Unscaled multiplicity
$\max_i$	Maximum over all values of i [e.g. if $x(1) = 4$ and $x(2) = 5$ them $\max_i$ $\{x(i)\} = 5$]
ml	Masking level in dB (at P, due to a single pure tone component at P')

ML	Masking level in dB (at P, due to *all* pure tone components P')
n	Harmonic number; harmonic template component number
p	Sound pressure
P	Pitch category (middle $C = C_4 = 48$, $C_3 = 36$, etc.)
P_n	Pitch category of nth harmonic
r	Pitch register (in octaves)
R_e	Mean result of a single experimental trial
$\bar{R}_e$	Mean result of all trials in an experiment
R_t	Unscaled, theoretical (calculated) result of a trial
$\bar{R}_t$	Mean of R_t over all trials in an experiment
R'_t	R_t scaled against R_e by linear transformation
s	Simultaneity number (in a progression)
S	Tone salience; salience of a tone sensation
S_p	Chroma probability (measure of chroma salience)
S_t	Chroma tally (measure of chroma salience)
SPL	Sound pressure level (in dB above 2×10^{-5} Pa)
T_c	Complex tonalness of a simultaneity
T_p	Pure tonalness of a simultaneity
TL	Threshold level (level at threshold of audibility); SPL in dB of an isolated, barely audible pure tone
W	Harmonic weight; importance of a harmonic template component
W_{cb}	Critical bandwidth (in semitones)
x	Arbitrary real variable
X	Apparent pitch proximity of two simultaneities
yl	Auditory level in dB of a pure tone component of a complex tone, which (in turn) is a component of a simultaneity (cf. al)
YL	Auditory level in dB of a pure tone component of a complex sound; level in dB above the threshold of audibility in quiet (cf. AL)
σ	Standard deviation
♭	Musical flat (lowers pitch by one semitone)
#	Musical sharp (raises pitch by one semitone)

References

Alwell, E., Schachter, C. (1978): *Harmony and Voice Leading 1.* (Harcourt Brace Jovanovich, New York)

Allen, D. (1967): Octave discriminability of musical and non-musical subjects. Psychonom. Sci. **7**, 421–422

Alphonce, B.H. (1980): Music analysis by computer – A field for theory formation. Comput. Music J. **4**(2), 26–35

American Standards Association (1960): *USA Standard Acoustical Terminology* (American Standards Association, New York)

Apel, W. (1970): *The Harvard Dictionary of Music*, 2nd. ed. (Harvard University Press, Cambridge, MA)

Armitage, S.E., Baldwin, B.A., Vince, M.A. (1980): The fetal sound environment of sheep. Science **208**, 1173–1174

Arnold, D., Fortune, N. (eds.) (1968): *The Monteverdi Companion* (Faber and Faber, London)

Attneave, F., Olsen, R.K. (1971): Pitch as a medium: A new approach to psychophysical scaling. Am. J. Psychol. **84**, 147–166

Aures, W. (1984): "Berechnungsverfahren für den Wohlklang beliebiger Schallsignale, ein Beitrag zur gehörbezogenen Schallanalyse"; Doctoral dissertation, Technical University of Munich

Averbach, E., Coriell, A.S. (1961): Short-term memory in vision. Bell Syst. Mech. J. **40**, 309–328. [Cited in Adams, J.A. (1980): *Learning and Memory: An Introduction* (Dorsey, Homewood, IL)]

Babbitt, M. (1955): Some aspects of twelve-tone composition. Score IMA Mag. **12** [Cited in Forte (1964)]

Bachem, A. (1948): Chroma fixation at the end of the musical frequency scale. J. Acoust. Soc. Am. **20**, 704–705

Backus, J. (1969): *The Acoustical Foundations of Music* (Norton and Murray, London)

Balzano, G.J. (1980): The group-theoretical description of 12-fold and microtonal pitch systems. Comput. Music J. **4**(4), 66–84

Balzano, G.J. (1982): "The Pitch Set as a Level of Description for Studying Musical Pitch Perception", in *Music, Mind and Brain: The Neuropsychology of Music*, ed. by M. Clynes (Plenum, New York)

Balzano, G.J. (1984): Absolute pitch and pure tone identification. J. Acoust. Soc. Am. **75**, 623–625

Barlow, C. (1987): Two essays on theory. Comput. Music J. **11**, 44–60

Beach, S. (1979): Pitch structure and the analytic process in atonal music: An interpretation of the theory of sets. Music Theo. Spectrum **1**, 7–22 [Cited in Gibson, D.B. (1986): J. Res. Music Educ. **34**, 5–23]

Beck, J., Shaw, W.A. (1963): Single estimates of pitch magnitude. J. Acoust. Soc. Am. **35**, 1722–1724

Bekesy, G. von (1947): The variation of phase along the basilar membrane with sinusoidal vibrations. J. Acoust. Soc. Am. **19**, 452–460

Benade, A. (1976): *Fundamentals of Musical Acoustics* (Oxford University Press, Oxford)

Bench, J. (1968): Sound transmission to the human foetus through the maternal abdominal wall. J. Genet. Psychol. **113**, 85–87 [Cited in Vince et al. (1982a)]

Beranek, L. L. (1972): "Acoustical Definitions", in *American Institute of Physics Handbook*, 3rd ed., ed. by D. E. Gray (McGraw-Hill, New York)

Berger, K. (1964): Some factors in the recognition of timbre. J. Acoust. Soc. Am. **36**, 1888 [Cited in Rasch (1979)]

Bernard, C. G., Kaiser, I. H., Kilmodin, G. M. (1959): On the development of cortical activity in fetal sheep. Acta Physiol. Scand. **47**, 333–349

Berry, W. (1976): *Structural Functions in Music* (Prentice Hall, Englewood Cliffs, NJ) [Cited in Erickson (1982)]

Bharucha, J., Krumhansl, C. L. (1983): The representation of harmonic structure in music: Hierarchies of stability as a function of context. Cognition **13**, 63–102

Bharucha, J. J., Stoeckig, K. (1987): Priming of chords: Spreading activation of overlapping frequency spectra? Percept. Psychophys. **41**, 519–524

Boomsliter, P., Creel, W. (1961): The long pattern hypothesis in harmony and hearing. J. Music Theo. **5**, 2–31

Booth, M. W. (1981): *The Experience of Song* (Yale University Press, New Haven, CT)

Boring, E. G. (1942): *Sensation and Perception in the History of Experimental Psychology* (Appleton–Century Crofts, New York)

Brady, P T. (1970): Fixed-scale mechanism of absolute pitch. J. Acoust. Soc. Am. **48**, 883–887

Bregman, A. S. (1981): "Asking the 'What for' Question in Auditory Perception", in *Perceptual Organization*, ed. by M. Kubovy, J. R. Pomerantz (Erlbaum, Hillsdale, NJ)

Bregman, A. S., Campbell, J. (1971): Primary auditory stream segregation and perception of order in rapid sequences of tone. J. Exp. Psychol. **89**, 244–249

Bregman, A. S., Pinker, S. (1978): Auditory streaming and the building of timbre. Can J. Psychol. **32**, 19–31

Bregman, A. S., Abramson, J., Doehring, P., Darwin, C. J. (1985): Spectral integration based on common amplitude modulation. Percept. Psychophys. **37**, 483–493

Brink, G. van den (1982): On the relativity of pitch. Perception **11**, 721–731

Browne, R. (1981): Tonal implications of the diatonic set. In Theory Only **5**, 3–21

Bruner, C. L. (1984): The perception of contemporary pitch structures. Music Percept. **2**, 25–40

Burns, E. M., Ward, W. D. (1978): Categorical perception – Phenomenon or epiphenomenon: Evidence from experiments in the perception of melodic musical intervals. J. Acoust. Soc. Am. **63**, 456–468

Burns, E. M., Ward, W. D. (1982): "Intervals, Scales and Tuning", in *The Psychology of Music*, ed. by D. Deutsch (Academic, New York) pp. 241–269

Butler, D., Brown, H. (1984): Tonal structure versus function: Studies in the recognition of harmonic motion. Music Percept. **1**, 6–24

Castellano, M. A., Bharucha, J., Krumhansl. C. L. (1984): Tonal hierarchies in the music of North India. J. Exp. Psychol.: Gen. **113**, 394–412

Cattell, J. M. (1886): The inertia of the eye and brain. Brain **8**, 295–312 [Cited in Sekuler, R., Blake, R. (1985): *Perception* (Knopf, New York)]

Cazden, N. (1954): Hindemith and nature. Music Rev. **15**, 288–306

Cazden, N. (1958): Pythagoras and Aristoxenos reconciled. J. Am. Musicol. Soc. **11**, 97–105

Cazden, N. (1972): The systematic references of musical consonance response. Int. Rev. Aesthetics Sociol. Music **3**, 217–235

Christensen, T. (1987): 18th-century science and the *corps sonore*: The scientific background to Rameau's principle of harmony. J. Music Theo. **31**, 22–50

Clarkson, M. G., Clifton, R. K. (1985): Infant pitch perception: Evidence for responding to pitch categories and the missing fundamental. J. Acoust. Soc. Am. **77**, 1521–1528

Clifton, T. (1983): *Music as Heard: A Study in Applied Phenomenology* (Yale University Press, New Haven, CT) [Reviewed by J. Tenney: J. Music Therapy **29**, 197–213 (1985)]

Clynes, M. (1977): *Sentics: The Touch of the Emotions* (Doubleday/Anchor, New York)

Clynes, M., Walker, J. (1982): "Rhythm, time and pulse in music: Neurobiologic functions", in *Music, Mind and Brain: The Neuropsychology of Music*, ed. by M. Clynes (Plenum, New York)

Cohen, A. J. (1962): Information theory and music. Behav. Sci. **7**, 137–163
Cohen, A. J. (1982): Exploring the sensitivity to structure in music. Can. Univ. Music Rev. **3**, 15–30
Cohen, A. J., Thorpe, L. A., Trehub, S. E. (1987): Infants' perception of musical relations in short transposed tone sequences. Can. J. Psychol. **41**, 33–47
Cohen, E. A. (1984): Some effects of inharmonic partials on interval perception. Music Percept. **1**, 323–349
Cooke, D. (1959): *The Language of Music* (Oxford University Press, Oxford)
Corso, J. F. (1954): Scale position and performed melodic octaves. J. Psychol. **37**, 297–305 [Cited in Terhardt (1978)]
Costall, A. (1985): "The Relativity of Absolute Pitch", in *Musical Structure and Cognition*, ed. by P. Howell, I. Cross, R. West (Academic, London)
Craik, F. I. M., Lockhart, R. S. (1972): Levels of processing: A framework for memory research. J. Verbal Learning Verbal Behav. **11**, 671–684 [Cited in Deutsch and Feroe (1981)]
Cross, I., West, R., Howell, P. (1985): Pitch relations and the formation of scalar structure. Music Percept. **2**, 329–344
Crowder, R. G. (1970): The role of one's own voice in immediate memory. Cognitive Psychol. **1**, 157–178 [Cited in Adams, J. A. (1980): *Learning and Memory: An Introduction* (Dorsey, Homewood, IL)]
Crowder, R. G. (1986): Perception of the major/minor distinction II: Experimental investigations. Psychomusicology **5**, 3–24
Crowder, R. G., Morton, J. (1969): Precategorical acoustic storage (P. A. S.). Percept. Psychophys. **5**, 365–373
Crowe, S. T., Guild, S. R., Polvost, L. M. (1934): Observations on the pathology of high-tone deafness. Bull. Johns Hopkins Hosp. **54**, 315–379 [Cited in Sekuler, R., Blake, R. (1985): *Perception* (Knopf, New York)]
Cuddy, L. L. (1968): Practice effects in the absolute judgement of pitch. J. Acoust. Soc. Am. **43**, 1069–1076
Cuddy, L. L., Badertscher, B. (1987): Recovery of the tonal hierarchy: Some comparisons across age and musical experience. Percept. Psychophys. **41**, 609–620
Cuddy, L. L., Lyons, H. I. (1981): Musical pattern recognition: A comparison of listening to and studying tonal structures and tonal ambiguities. Psychomusicology **1**, 15–33
Cuddy, L. L., Cohen, A. J., Miller, J. (1979): Melody recognition: The experimental application of musical rules. Can J. Psychol. **33**, 148–157

Dalglish, W. (1978): The origin of the hocket. J. Am. Musicol. Soc. **31**, 3–20
Davidson, B., Power, R. P., Michi, P. T. (1987): The effects of familiarity and previous training on perception of an ambiguous musical figure. Percept. Psychophys. **41**, 601–608
Davidson, L. (1985): Tonal structures of children's early songs. Music Percept. **2**, 361–374
Davies, J. (1979): Memory for melodies and tonal sequences: A theoretical note. Br. J. Psychol. **70**, 205–210
De Klerk, D. (1979): Equal temperament. Acta Musicol. **51**, 140–150
De la Motte-Haber, H. (1985): "Rationalität und Affect – über das Verhältnis von mathematischer Begründung und psychologischer Wirkung der Musik", in *Musik und Mathematik*, ed. by H. Götze, R. Wille (Springer, Berlin, Heidelberg)
Deliege, I. (1987): Grouping conditions in listening to music: An approach to Lerdahl and Jackendoff's grouping preference rules. Music Percept. **4**, 325–360
Demany, L. (1982): Auditory stream segregation in infancy. Infant Behav. Dev. **5**, 261–276 [Cited in Trehub (1987)]
Demany, L., Armand, F. (1984) The perceptual reality of tone chroma in early infancy. J. Acoust. Soc. Am. **76**, 57–66
Deutsch, D. (1972a): Effect of repetition of standard and comparison tones in recognition memory for pitch. J. Exp. Psychol. **93**, 156–162
Deutsch, D. (1972b): Octave generalization and tune recognition. Percept. Psychophys. **11**, 411–412

Deutsch, D. (1973): Octave generalization of specific interference effects in memory for tonal pitch. Percept. Psychophys. **13**, 271–275

Deutsch, D. (1982a): "The Processing of Pitch Combinations", in *The Psychology in Music*, ed. by D. Deutsch (Academic, New York)

Deutsch, D. (1982b): "Organizational Processes in Music", in *Music, Mind and Brain: The Neuropsychology of Music*, ed. by M. Clynes (Plenum, New York)

Deutsch, D. (1987): The tritone paradox: Effects of spectral variables. Percept. Psychophys. **41**, 563–575

Deutsch, D., Boulanger, R.C. (1984): Octave equivalence and the immediate recall of pitch sequences. Music Percept. **2**, 40–51

Deutsch, D., Feroe, J. (1981): The internal representation of pitch sequences in tonal music. Psychol. Rev. **88**, 503–522

Deutsch, D., Moore, F.R., Dolson, M. (1984): Pitch classes differ with respect to height. Music Percept. **2**, 265–271

Dewar, K.M., Cuddy, L.L., Mewhort, D.J.K. (1977): Recognition memory for single tones with and without context. J. Exp. Psychol.: Human Learning and Memory **3**, 60–67

DeWitt, L.A., Crowder, R.G. (1987): Tonal fusion of consonant musical intervals: The Oomph in Stumpf. Percept. Psychophys. **41**, 73–84

Dowling, W.J. (1973): The perception of interleaved melodies. Cognitive Psychol. **5**, 332–337

Dowling, W.J. (1978): Scale and contour: Two components of a theory of memory for melodies. Psychol. Rev. **85**, 341–354

Dowling, W.J. (1982): "Melodic Information Processing and Its Development", in *The Psychology of Music*, ed. by D. Deutsch (Academic, New York)

Dowling, W.J., Fujitani, D.A. (1971): Contour, interval and pitch recognition in memory for melodies. J. Acoust. Soc. Am. **49**, 524–531

Dowling, W.J., Hollombe, A.W. (1977): The perception of melodies distorted by splitting into several octaves: Effects of increasing proximity and melodic contour. Percept. Psychophys. **21**, 60–64

Dowling, W.J., Lung, K., Herbold, S. (1987): Aiming attention in pitch and time in the perception of interleaved melodies. Percept. Psychophys. **41**, 642–656

Drever, J. (1964): *Penguin Dictionary of Psychology* (Penguin, London)

Duke, R.A. (1985): Wind instrumentalists' intonation performance of selected musical intervals. J. Res. Music Educ. **33**, 101–111

Edworthy, J. (1985): Interval and contour in melody processing. Music Percept. **2**, 375–388

Egan, J.P., Klumpp, R.G. (1951): The error due to masking in the measurement of aural harmonics by the method of best beats. J. Acoust. Soc. Am. **23**, 275–286 [Cited in Ward (1970)]

Egan, J.P., Meyer, D.R. (1950): Changes in pitch of tones of low frequency as a function of the pattern of excitation produced by a band of noise. J. Acoust. Soc. Am. **22**, 827–833 [Cited in Terhardt (1979a)]

Eimas, P.D., Siqueland, E.R., Jusczyk, P., Vigoroto, J.M. (1971): Speech perception in infants. Science **71**, 303–306

Eiting, M.H. (1984): Perceptual similarities between musical motifs. Music Percept. **2**, 78–94

Ellis, C.J. (1967): The Pitjantjara Kangaroo Song from Karla. Misc. Musicolog. **2**, 171–247

Ellis, C.J. (1985): *Aboriginal Music, Education for Living* (University of Queensland Press, St. Lucia, Queensland)

Elmasian, R., Birnbaum, M.H. (1984): A harmonious note on pitch: Scales of pitch derived from subtractive model of comparison agree with the musical scale. Percept. Psychophys. **36**, 531–537

Erickson, R. (1982): "New Music and Psychology", in *The Psychology of Music*, ed. by D. Deutsch (Academic, New York)

Erickson, R. (1984): A perceptual substrate for tonal centering? Music Percept. **2**, 1–5

Eriksen, C.W., Johnson, H.J. (1964): Storage and decay characteristics of non-attended auditory stimuli. J. Exp. Psychol. **68**, 28–36 [Cited in Neisser (1967)]

Evans, E.F. (1975): The sharpening of cochlear frequency selectivity in the normal and abnormal cochlea. Audiology **14**, 419–442 [Cited in Moore (1982)]

Fastl, H. (1971): Über Tonhöhenempfindung bei Rauschen. Acustica **25**, 350–354
Fastl, H., Hesse, A. (1984): Frequency discrimination for pure tones at short durations. Acustica **56**, 41–47
Fastl, H., Stoll, G. (1979): Scaling of pitch strength. Hearing Res. **1**, 293–301
Fechner, G.T. (1860): *Elemente der Psychophysik* (Breitkopf und Härtel, Leipzig)
Feld, S. (1974): Linguistic models in ethnomusicology. Ethnomusicology **18**, 197–217
Fidell, S., Horonjeff, R., Teffeteller, S., Green, D.M. (1983): Effective masking bandwidth at low frequencies. J. Acoust. Soc. Am. **73**, 628–638
Flanagan, J.L., Guttman, N. (1960): On the pitch of periodic pulses. J. Acoust. Soc. Am. **32**, 1308–1319 [Cited in Terhardt et al. (1982a)]
Fletcher, H. (1924): The physical criterion for determining the pitch of a musical tone. Phys. Rev. **23**, 427–437 [Cited in Plomp (1967)]
Fletcher, H. (1934): Loudness, pitch and timbre of musical tones and their relations to the intensity, the frequency and the overtone structure. J Acoust. Soc. Am. **6**, 59–69
Fletcher, H. (1940): Auditory patterns. Rev. Mod. Phys. **12**, 47–65 [Cited in Moore (1982)]
Fletcher, H., Galt, R.H. (1950): The perception of speech and its relation to telephony. J. Acoust. Soc. Am. **22**, 89–151 [Cited in Terhardt et al. (1982b)]
Fletcher, H., Munson, W.A. (1933): Loudness, its definition, measurement and calculation. J. Acoust. Soc. Am. **5**, 82
Fokker, A.D. (1966): On the expansion of the musician's realm of harmony. Acta Musicol. **38**, 197–202
Forte, A. (1962): *Tonal Harmony in Concept and Practice* (Holt, Rinehardt and Winston, New York)
Forte, A. (1964): A theory of set-complexes for music. J. Music Theo. **8**, 136–193
Forte, A. (1973): *The structure of atonal music* (Yale University Press, New Haven, CT) [Cited in Gibson, D.B. (1986): J. Res. Music Educ. **34**, 5–23]
Francès, R. (1972): *La Perception de la Musique*, 2nd ed. (Vrin, Paris) [Cited in Francès, R. (1985): Tonal principles as teaching principles in music. Music Percept. **2**, 389–396]
Fransson, F., Sundberg, J., Tjernlund, P. (1974): The scale in played music. Swed. J. Musicol. **56**, 49–54

Gabrielsson, A., Bengtsson, I., Gabrielsson, B. (1983): Performance of musical rhythm in 3/4 and 6/8 meter. Scand. J. Psychol. **24**, 193–213
Galliver, D. (1969): "Favolare in armonia" – A speculation into aspects of 17th century singing. Misc. Musicol. **4**, 129–146
Gardner, A.D., Pickford, R.W. (1943, 1944): Relation between dissonance and context. Nature **152**, 358; **154**, 274–275 [Cited in Lundin (1947)]
Garner, W.R. (1970): Good patterns have few alternatives. Am. Sci. **58**, 34–42
Gfeller, K.E. (1983): Musical mnemonics as an aid to retention with normal and learning disabled students. J. Music Therapy **20**, 179–189
Gibson, E.J. (1953): Improvement in perceptual judgments as a function of controlled practice or training. Psychol. Bull. **50**, 401–431
Gibson, J.J. (1966): *The Senses Considered as Perceptual Systems* (Houghton Mifflin, Boston)
Gibson, J.J. (1979): *The Ecological Approach to Visual Perception* (Houghton Mifflin, Boston)
Glucksberg, S., Cowen, G.N., Jr. (1970): Memory for nonattended auditory material. Cognitive Psychol. **1**, 149–156
Goldman, R. (1965): *Harmony in Western Music* (Barrie and Rockliff, London)
Goldstein, J.L. (1973): An optimum processor theory for the central formation of the pitch of complex tones. J. Acoust. Soc. Am. **54**, 1496–1516
Grant, R.E., LeCroy, S. (1986): Effects of sensory mode input on the performance of rhythmic perception tasks by mentally retarded subjects. J. Music Therapy **23**, 2–9
Greene, P.C. (1937): Violin intonation. J. Acoust. Soc. Am. **9**, 43–44 [Cited in Ward (1970)]

Grey, J.M. (1977): Multidimensional perceptual scaling of musical timbres. J. Acoust. Soc. Am. **61**, 1270

Griffiths, P. (1978): *A Concise History of Modern Music from Debussy to Boulez* (Thames and Hudson, London)

Grimwade. J.C., Walker, D.W., Wood, C. (1970): Sensory stimulation of the human foetus. Aust. J. Mental Retard. **2**, 63–64

Grout, D.J. (1960): *A History of Western Music* (Norton, New York)

Haase, R. (1986): Harmonikale Grundlagenforschung. Acta Musicol. **58**, 282–304

Hagerman, B., Sundberg, J. (1980): Fundamental frequency adjustment in barbershop singing. J. Res. Singing **4**, 3–17

Hall, D.E. (1980): *Musical Acoustics: An Introduction* (Wadsworth, Belmont, CA)

Hall, D.E., Hess, J.T. (1984): Perception of musical interval tuning. Music Percept. **2**, 166–195

Hall, J.W., III, Peters, R.W. (1981): Pitch for nonsimultaneous successive harmonics in quiet and noise. J. Acoust. Soc. Am. **69**, 509–513

Hanser, S.B. (1985): Music therapy and stress reduction research. J. Music Therapy **22**, 193–206

Harris, G., Siegel, J.A. (1975): Categorical perception and absolute pitch. J. Acoust. Soc. Am. **57**, S11

Harwood, D.L. (1976): Universals in music: A perspective from cognitive psychology. Ethnomusicology **20**, 521–533

Heinbach, W. (1986): Untersuchung einer gehörbezogenen Spektralanalyse mittels Resynthese. Fortschr. Akustik – DAGA 453–456

Helmholtz, H.L.F. von (1863), English transl. A.J. Ellis (1954): *On the Sensations of Tone as a Physiological Basis for the Theory of Music* (Dover, New York)

Hesse, A. (1985): Zur Ausgeprägtheit der Tonhöhe gedrosselter Sinustöne. Fortschr. Akustik – DAGA 535–538

Hesse, A. (1987): Ein Funktionsschema der Spektraltonhöhe von Sinustönen. Acustica **63**, 1–16

Hesse, H.P. (1982): "The Judgment of Musical Intervals", in *Music, Mind and Brain: The Neuropsychology of Music*, ed. by M. Clynes (Plenum, New York)

Heyde, E.M. (1987): *Was ist absolutes Hören? Eine musikpsychologische Untersuchung* (Profil, Munich) [Reviewed by T.H. Stoffer: *Neue Zeitung für Musik* **148**/6, 56 (1987)]

Heyduk, R.G. (1975): Rated preference for musical compositions as it relates to complexity and exposure frequency. Percept. Psychophys. **17**, 84–91

Hiller, L.A., Isaacson, L.M. (1959): *Experimental Music* (McGraw-Hill, New York)

Hindemith, P. (1940), English transl. A. Mendel (1942): *The Craft of Musical Composition* (Schott, Mainz)

Hoerner, S. von (1976): The definition of major scales for chromatic scales of 12, 19 and 31 divisions per octave. Psychol. Music **4**, 13–23

Hofstadter, D.R. (1980): *Gödel, Escher, Bach: An Eternal Golden Braid* (Penguin, New York)

Houtgast, T. (1976): Subharmonic pitches of a pure tone at low S/N ratio. J. Acoust. Soc. Am. **60**, 405–409

Houtgast, T. (1977): Auditory filter characteristics derived from direct masking data and pulsation threshold data with a rippled-noise masker. J. Acoust. Soc. Am. **62**, 409–415

Houtsma, A.J.M. (1979): Musical pitch of two-tone complexes and predictions by modern pitch theories. J. Acoust. Soc. Am. **66**, 87–99

Houtsma, A.J.M., Rossing, T.D. (1987): Effects of signal envelope on the pitch of short complex tones. J. Acoust. Soc. Am. **81**, 439–444

Hutchinson, W., Knopoff, L. (1978): The acoustic component of Western consonance. Interface **7**, 1–29

Idson, W.L., Massaro, D.W. (1978): A bidimensional model of pitch in the recognition of melodies. Percept. Psychophys. **24**, 551–565

Jeffries, T.B. (1972): Familarity-frequency ratings of melodic intervals. J. Res. Music Educ. **20**, 391–396

Jeffries, T.B. (1974): Relationship of interval frequency count to ratings of melodic intervals. J. Exp. Psychol. **102**, 903–905
Jhairazboy, N., Stone, A. (1963): Intonation in present day North Indian classical music. Bull. School Oriental African Stud. (Univ. London) **26**, 118–132 [Cited in Burns and Ward (1978)]
Jones, M.R. (1987): Dynamic pattern structure in music: Recent theory and research. Percept. Psychophys. **41**, 621–634
Jordan, D.S. (1987): Influence of the diatonic tonal hierarchy at microtonal intervals. Percept. Psychophys. **41**, 482–488
Jordan, D.S., Shepard, R.N. (1987): Tonal schemas: Evidence obtained by probing distorted musical scales. Percept Psychophys. **41**, 489–504
Jorgenson, D. (1963): A resumé of harmonic dualism. Music Lett. **44**, 31–42
Juni, S., Nelson, S.P., Brannon, R. (1987): Minor tonality music preference and oral dependency. J. Psychol. **121**, 229–236

Kallman, H.J. (1982): Octave equivalence as measured by similarity ratings. Percept. Psychophys. **32**, 37–49
Kallman, H.J., Massaro, D.W. (1979): Tone chroma is functional in melody recognition. Percept. Psychophys. **26**, 32–36
Karp, C. (1984): A matrix technique for analyzing musical tuning systems. Acustica **54**, 209–216
Kennan, K.W. (1970): *The Technique of Orchestration*, 2nd ed. (Prentice-Hall, Englewood Cliffs, NJ)
Kessler, E.J., Hansen, C., Shepard, R.N. (1984): Tonal schemata in the perception of music in Bali and in the West. Music Percept. **2**, 131–165
Klingenberg, H.G. (1974): Grenzen der akustischen Gedächtnisfähigkeit. Acta Musicol. **46**, 171–180
Knupp, R. (1984): The tonal structure of *Tristan und Isolde*: A sketch. Music Rev. **45**, 11–25
Koffka, K. (1935): *Principles of Gestalt Psychology* (Harcourt, Brace and World, New York)
König, E. (1957): Effect of time on pitch discrimination thresholds under several psychophysical procedures; comparison with intensity discrimination thresholds. J. Acoust. Soc. Am. **29**, 606–612 [Cited in Ward (1970)]
Kowal, K. (1987): Apparent duration and numerosity as a function of melodic familiarity. Percept. Psychophys. **42**, 122–131
Kronman, U., Sundberg, J. (1987): "Is the musical ritard an illusion to physical motion?" in *Action and Perception in Rhythm and Music*, ed. by A. Gabrielsson (Royal Swedish Academy of Music, Stockholm)
Krueger, F. (1910): Die Theorie der Konsonanz. Psychol. Studien **5**, 294–411 [Cited in Cazden (1954)]
Krumhansl, C.L. (1979): The psychological representation of musical pitch in a tonal context. Cognitive Psychol. **11**, 346–374
Krumhansl, C.L. (1983): Perceptual structures for tonal music. Music Percept. **1**, 28–62
Krumhansl, C.L., Castellano, M.A. (1983): Dynamic processes in music perception. Memory Cognition **11**, 325–334
Krumhansl, C.L., Kessler, E.J. (1982): Tracing the dynamic changes in perceived tonal organization in a spatial representation of musical keys. Psychol. Rev. **89**, 334–368
Krumhansl, C.L., Shepard, R.N. (1979): Quantification of the hierarchy of tonal functions within a diatonic context. J. Exp. Psychol.: Human Percept. Perform. **5**, 579–594
Krumhansl, C.L., Bharucha, J., Kessler, E.J. (1982): Perceived harmonic structure of chords in three related musical keys. J. Exp. Psychol.: Human Percept. Perform. **8**, 24–36
Kruskal, J.B. (1964): Nonmetric multidimensional scaling: A numerical method. Psychometrika **29**, 28–42 [Cited in Krumhansl and Kessler (1982)]
Kubovy, M. (1981): "Concurrent-Pitch Segregation and the Theory of Indispensible Attributes", in *Perceptual Organization*, ed. by M. Kubovy, J.R. Pomeranz (Erlbaum, Hillsdale, NJ) [Cited in Sloboda (1985)]
Kuhn, T.S. (1962): *The Structure of Scientific Revolutions* (Chicago University Press, Chicago)

Landau, V. (1961): Hindemith the system builder: A critique of his theory of harmony. Music Rev. **22**, 136–151

Lehr, A. (1986): Partial groups in the bell sound. J. Acoust. Soc. Am. **79**, 2000–2011

Lerdahl, F., Jackendoff, R. (1977): Toward a formal theory of tonal music. J. Music Theo. **21**, 110–171

Lerdahl, F., Jackendoff, R. (1983): *A Generative Theory of Tonal Music* (MIT Press, Cambridge, MA)

Lewis, D., Cowan, M. (1936): The influence of intensity on the pitch of violin and 'cello tones. J. Acoust. Soc. Am. **8**, 20–22 [Cited in Ward (1970)]

Liberman, A.M., Harris, K.S., Hoffman, H.S., Griffith, B.S. (1957): The discrimination of speech sounds within and across phoneme boundaries. J. Exp. Psychol. **54**, 358–368

Liberman, A.M., Harris, K.S., Kinney, J.A., Lane, H. (1961): The discrimination of relative onset-time of the components of certain speech and nonspeech patterns. J. Exp. Psychol. **61**, 379–388

Lieberman, P., Michaels, S.B. (1962): Some aspects of fundamental frequency and envelope amplitude as related to the emotional content of speech. J. Acoust. Soc. Am. **34**, 922–927

Lilly, J.C. (1974): *The Human Biocomputer* (Abacus, London)

Lippman, E.A. (1963a): Hellenic conceptions of harmony. J. Am. Musicol. Soc. **16**, 3–35

Lippman, E.A. (1963b): Spatial perception and physical location as factors in music. Acta Musicol. **35**, 24–34

Lloyd, L.S. (1954): The strike-notes of church bells. Music Lett. **35**, 227–232, 240

Lockard, J.S., Daley, P.C., Gunderson, V.M. (1979): Maternal and paternal differences in infant carry: US and African data. Am. Nat. **113**, 235–246

Lockhead, G.R., Byrd, R. (1981): Practically perfect pitch. J. Acoust. Soc. Am. **70**, 387–389

Longuet-Higgins, H.C. (1979): The perception of music. Proc. R. Soc. London B **205**, 307–322

Lufti, R.A. (1985): A power-law transformation predicting masking by sounds with complex spectra. J. Acoust. Soc. Am. **77**, 2128–2136

Lundin, R.W. (1947): Toward a cultural theory of consonance. J. Psychol. **23**, 45–49

MacPherson, S. (1920): *Melody and Harmony* (Williams, London)

Madsen, C.K., Geringer, J.M. (1981): Discrimination between tone quality and intonation in unaccompanied flute/oboe duets. J. Res. Music Educ. **29**, 305–313

Maher, T.F. (1980): A rigorous test of the proposition that musical intervals have different psychological effects. Am. J. Psychol. **93**, 309–327

Malmberg, C.F. (1918): The perception of consonance and dissonance. Psychol. Monogr. **25**, 93–133

Massaro, D.W. (1970): Retroactive interference in short-term recognition memory of pitch. J. Exp. Psychol. **83**, 32–39

Massaro, D.W., Kallman, H.J., Kelly, J.L. (1980): The role of tone height, melodic contour and tone chroma in melody recognition. J. Exp. Psychol.: Human Learning Memory **6**, 77–90

Matthews, M.V., Pierce, J.R. (1980): Harmony and non-harmonic partials. J. Acoust, Soc. Am. **68**, 1252–1257

McAdams, S. (1984): "The Auditory Image: A Metaphor for Musical and Psychological Research on Auditory Organization", in *Cognitive Processes in the Perception of Art*, ed. by W.R. Crozier, A.J. Chapman (Elsevier, Amsterdam)

McAdams, S., Saariaho, K. (1985): Qualities and functions of musical timbre. Proc. 1985 Int. Computer Music Conf., ed. by B. Truax (Computer Music Association, San Francisco)

McClain, E.G. (1979): Chinese cyclic tunings in late antiquity. Ethnomusicology **23**, 205–224

Meyer, J. (1979): Zur Tonhöhenempfindung bei musikalischen Klängen in Abhängigkeit vom Grad der Gehörschulung. Acustica **42**, 189–204

Meyer, L.B. (1956): *Emotion and Meaning in Music* (University of Chicago Press, Chicago)

Meyer, L.B. (1973): *Explaining Music: Essays and Explorations* (University of California Press, Berkeley, CA)

Meyer, M.F. (1962): Helmholtz's aversion to tempered tuning experimentally shown to be a neurological problem. J. Acoust. Soc. Am. **34**, 127–128

Miller, G. A. (1956): The magic number seven, plus or minus two: Some limitations on our capacity for processing information. Psychol. Rev. **63**, 81–97

Miller, G. A., Heise, G. A. (1950): The trill threshold. J. Acoust. Soc. Am. **22**, 637–638

Miller, G. A., Licklider, J. C. R. (1950): The intelligibility of interrupted speech. J. Acoust. Soc. Am. **22**, 167–173

Mitchell, W. J. (1967): The Tristan prelude: Techniques and structure. Music Forum **1**, 162–203

Monahan, C. B., Kendall, R. A., Carterette, E. C. (1987): The effect of melodic and temporal contour on recognition memory for pitch change. Percept. Psychophys. **41**, 576–600

Moore, B. C. J. (1982): *An Introduction to the Psychology of Hearing*, 2nd ed. (Academic, London)

Moore, B. C. J. (1985): Additivity of simultaneous masking, revisited. J. Acoust. Soc. Am. **78**, 488–494

Moore, B. C. J., Peters, R. W., Glasberg, B. R. (1985): Thresholds for the detection of inharmonicity in complex tones. J. Acoust. Soc. Am. **77**, 1861–1867

Moore, B. C. J., Rosen, S. M. (1979): Tune recognition with reduced pitch and interval information. Q. J. Exp. Psychol. **31**, 229–240

Moran, H., Pratt, C. C. (1926): Variability of judgments of musical intervals. J. Exp. Psychol. **9**, 492–500 [Cited in Burns and Ward (1978)]

Morgan, C. T., Garner, W. R., Galambos, R. (1951): Pitch and intensity. J. Acoust. Soc. Am. **23**, 658–663 [Cited in Ward (1970)]

Morris, G. (1986): "Reverse Cauchy Inequality"; Manuscript available from the Department of Mathematics, University of New England, Armidale NSW, Australia

Nakamura, T. (1987): The communication of dynamics between musicians and listeners through musical performance. Percept. Psychophys. **41**, 525–533

Narmour, E. (1983): Some major theoretical problems concerning the concept of hierarchy in the analysis of tonal music. Music Percept. **1**, 129–199

Neisser, U. (1967): *Cognitive Psychology* (Meredith, New York)

Nettl, B. (1956): *Music in Primitive Culture* (Harvard University Press, Cambridge, MA [Cited in Demany and Armand (1984)]

Nettl, B. (1986): World music in the twentieth century: A survey of research on Western influence. Acta Musicol. **58**, 360–373

Nickerson, J. F. (1949): Intonation of solo and ensemble performance of the same melody. J. Acoust. Soc. Am. **21**, 593–595

Nielsen, F. V. (1983): *Oplevelse af musikalsk spaending* (The experience of musical tension) (Akademisk Forlag, Copenhagen)

Nilsonne, A., Sundberg, J. (1984): Differences in the ability of musicians and nonmusicians to judge emotional state from the fundamental frequency of voice samples. Music Percept. **2**, 507–516

Noorden, L. van (1975): *Temporal Coherence in the Perception of Tone Sequences* (Institute for Perception Research, Eindhoven, The Netherlands)

Norden, N. L. (1936): A new theory of untempered music. Musical Q. **22**, 211–233 [Cited in Nickerson (1949)]

O'Connor, J. D., Arnold, G. F. (1973): *The Intonation of Colloquial English*, 2nd ed. (Longman, London)

Ohgushi, K. (1983): The origin of tonality and a possible explanation of the octave enlargement phenomenon. J. Acoust. Soc. Am. **73**, 1694–1700

Ohm, G. S. (1843): Über die Definition des Tones, nebst daran geknüpfter Theorie der Sirene und ähnlicher tonbildender Vorrichtungen. Ann. der Phys. Chem. **59**, 513–565

O'Keefe, V. (1975): Psychological preference for harmonized musical passages in the just and equally-tempered systems. Percept. Mot. Skills **40**, 192–194

Olsen, R. K., Hanson, V. (1977): Interference effects in tone memory. Memory Cognition **5**, 32–40

Parncutt, R. (1987 a). "Sensory Bases of Harmony in Western Music"; PhD thesis at the University of New England, Armidale NSW, Australia

Parncutt, R. (1987b): "The Perception of Pulse in Musical Rhythm", in *Action and Perception in Rhythm and Music*, ed. by A. Gabrielsson (Royal Swedish Academy of Music, Stockholm)

Patterson, R.D. (1976): Auditory filter shapes derived with noise stimuli. J. Acoust. Soc. Am. **59**, 640–654

Peacock, K. (1984): Synaesthetic perception: Alexander Skriabin's colour hearing. Music Percept. **2**, 483–506

Pierce, J.R. (1966): Attaining consonance in arbitrary scales. J. Acoust. Soc. Am. **40**, 249

Pierce, J.R. (1983): *The Science of Musical Sound* (Scientific American Library, New York)

Pikler, A.G. (1966): Logarithmic frequency systems. J. Acoust. Soc. Am. **39**, 1102–1110

Pipping, H. (1895): Zur Lehre von den Vocalklängen. Z. Biol. **13**, 524–583 [Cited in Plomp (1967)]

Piston, W. (1978): *Harmony*, 4th ed. (Norton, New York)

Plomp, R. (1964): The ear as frequency analyser I. J. Acoust. Soc. Am. **36**, 1628–1636

Plomp, R. (1965): Detectability threshold for combination tones. J. Acoust. Soc. Am. **37**, 1110–1123

Plomp, R. (1967): Pitch of complex tones. J. Acoust. Soc. Am. **41**, 1526–1533

Plomp, R., Levelt, W.J.M. (1965): Tonal consonance and critical bandwidth. J. Acoust. Soc. Am. **38**, 548–560

Plomp, R. Wagenaar, W.A., Mimpen, A.M. (1973): Musical interval recognition with simultaneous tones. Acustica **29**, 101–109

Popper, K.R. (1972): *Objective Knowledge: An Evolutionary Approach* (Oxford University Press, Oxford)

Popper, K.R., Eccles, J.C. (1977): *The Self and Its Brain* (Springer, Berlin, Heidelberg)

Pressing, J. (1978): Towards an understanding of scales in jazz. Jazz Res. **9**, 25–35

Rahn, J. (1980): On some computational models of music theory. Comput. Music J. **4** (2), 66–72

Rakowski, A. (1976): Tuning of isolated musical intervals. J. Acoust. Soc. Am. **59**, S50(A)

Rameau, J.-P. (1721), English transl. P. Gossett (1971): *Treatise on Harmony* (Dover, New York)

Rameau, J.-P. (1726): Nouveau système de musique théorique (Ballard, Paris) [Cited in Christensen (1987)]

Randel, D.M. (1971): Emerging triadic tonality in the fifteenth century. Musical Q. **57**, 73–86

Rasch, R.A. (1978): The Perception of simultaneous notes such as in polyphonic music. Acustica **40**, 21–33

Rasch, R.A. (1979): Synchronization in performed ensemble music. Acustica **43**, 119–131

Rasch, R.A. (1983): Description of regular 12-tone musical tunings. J. Acoust. Soc. Am. **73** 1023–1035

Rasch, R.A. (1985): Perception of melodic and harmonic intonation of two-part musical fragments. Music Percept. **2**, 441–458

Rasch, R.A., Plomp, R. (1982): "The Perception of Musical Tones" in *The Psychology of Music*, ed. by D. Deutsch (Academic, New York)

Read, G. (1987): *Sourcebook of Proposed Music Notation Reforms* (Greenwood, New York)

Révész, G. (1953): *Introduction to the Psychology of Music* (Longman, London)

Riemann, H. (1893): *Vereinfachte Harmonielehre* (Augener, London) [Cited in Apel (1970)]

Riesz, R.R. (1928): Differential intensity sensitivity of the ear for pure tones. Phys. Rev. **31**, 867–875

Risset, J.C. (1978): "Musical Acoustics", in *Handbook of Perception*, Vol. 4, ed. by E.C. Carterette, M.P. Friedman (Academic, New York) pp. 521–564

Ritsma, R.J. (1967): Frequencies dominant in the perception of the pitch of complex tones. J. Acoust. Soc. Am. **42**, 191–198

Roberts, L.A. (1986): Consonance judgements of musical chords by musicians and untrained listeners. Acustica **62**, 163–171

Roberts, L.A., Matthews, M.V. (1985): Intonation sensitivity for traditional and nontraditional chords. J. Acoust. Soc. Am. **75**, 952–959

Roederer, J.G. (1979): *Introduction to the Physics and Psychophysics of Music*, 2nd ed. (Springer, Berlin, Heidelberg)

Roederer, J.G. (1984): The search for a survival value of music. Music Percept. **1**, 350–356

Roederer, J.G. (1987): "Why Do We Love Music? A Search for the Survival Value of Music", in *Music in Medicine,* ed. by R. Spintge, R. Droh (Springer, Berlin, Heidelberg)

Rogers, G.L. (1987): Four cases of pitch-specific chromesthesia in trained musicians with absolute pitch. Psychol. Music **15**, 198–207

Rose, J.E., Brugge, J.F., Anderson, D.J., Hind, J.E. (1967): Phase-locked response to low-frequency tones in single auditory nerve fibers of the squirrel monkey. J. Neurophysiol. **30**, 769–793 [Cited in Sekuler, R., Blake, R. (1985): *Perception* (Knopf, New York)]

Rossing, T.D. (1982): *The Science of Sound* (Addison-Wesley, Reading, MA)

Ruckmick, C.A. (1929): A new classification of tonal qualities. Psychol. Rev. **36**, 172–180 [Cited in Shepard (1964)]

Saldanha, E., Corso, J. (1964): Timbre cues and the identification of musical instruments. J. Acoust. Soc. Am. **36**, 2021 [Cited in Rasch (1979)]

Salk, L. (1962): Mother's heart beat as an imprinting stimulus. Trans. NY Acad. Sci. **24**, 753 [Cited in Grimwade et al. (1970)]

Scharf, B. (1970): "Critical Bands", in *Foundations of Modern Auditory Theory*, ed. by J.V. Tobias (Academic, New York)

Schavernoch, H. (1981): *Die Harmonie der Sphären. Die Geschichte der Idee des Welteinklangs und der Seeleneinstimmung* (Arber, Freiburg) [Reviewed by E. Hickman: Musikforschung **40**, 81–82 (1987)]

Schenker, H. (1906), English transl. E.M. Borgese (1954): *Harmony* (University of Chicago Press, Chicago)

Schenker, H. (1935), English transl. E. Oster (1979): *Free Composition* (Longman, New York)

Schneider, M. (1960): Die musikalischen Grundlagen der Sphärenharmonie. Acta Musicol. **32**, 136–151

Schoenberg, A. (1911), English transl. R.E. Carter (1978): *Theory of Harmony* (Faber, London)

Schoenberg, A. (1954), ed. by L. Stein (1969): *Structural Functions of Harmony* (Norton, New York)

Scholes, P. (1970): *The Oxford Companion to Music*, 10th ed. (Oxford University Press, Oxford)

Schouten, J.F. (1938): The perception of subjective tones. Proc., K. Ned. Akad. Wet., **41**, 1086–1093

Schouten, J.F. (1940). The residue, a new component in subjective sound analysis. Proc., K. Ned. Akad. Wet. **43**, 356–365

Schuck, O.H., Young, R.W. (1943): Observations on the vibrations of piano strings. J. Acoust. Soc. Am. **15**, 1–11

Seashore, C.E. (1938): *Psychology of Music* (McGraw-Hill, New York)

Senju, M., Ohgushi, K. (1987): How are the player's ideas conveyed to the audience? Music Percept. **4**, 311–324

Sergeant, D. (1969): Experimental investigation of absolute pitch. J. Res. Music Educ. **17**, 135–143 [Cited in Sloboda (1985)]

Shackford, C. (1961, 1962): Some aspects of perception. J. Music Theo. **5**, 162–202; **6**, 66–90, 295–303 [Cited in Sundberg (1982)]

Shepard, R.N. (1964): Circularity in judgments of relative pitch. J. Acoust. Soc. Am. **36**, 2346–2353

Shepard, R.N. (1978): "The Circumplex and Related Topological Manifolds in the Study of Perception", in *Theory Construction and Data Analysis in the Behavioural Sciences,* ed. by S. Skye (Jessey-Bass, San Francisco) [Cited in Shepard (1982)]

Shepard, R.N. (1982): Geometrical approximations to the structure of musical pitch. Psychol. Rev. **89**, 305–333

Shirlaw, M. (1957): The science of harmony. Music Rev. **18**, 265–278

Siegel, J.A., Siegel, W. (1977a): Absolute identification of notes and intervals by musicians. Percept. Psychophys. **21**, 143–152

Siegel, J.A., Siegel, W. (1977b): Categorical perception of tonal intervals: Musicians can't tell sharp from flat. Percept. Psychophys. **21**, 399–407

Simmons, F.B., Edley, J.M., Lummins, R.C., Guttman, N., Frishkopf, L.S., Harmon, L.D., Zwicker, E. (1965): Auditory nerve: Electrical stimulation in man. Science **148**, 104–106

Singh, P.G. (1987): Perceptual organization of complex-tone sequences: A trade-off between pitch and timbre? J. Acoust. Soc. Am. **82**, 886–895

Sloboda, J.A. (1976): The effect of item position on the likelyhood of identification by inference in prose reading and music reading. Can. J. Psychol. **30**, 228–237

Sloboda, J.A. (1985): *The Musical Mind: The Cognitive Psychology of Music* (Clarendon, Oxford)

Smith, K.C., Cuddy, L.L. (1986): The pleasingness of melodic sequences: Contrasting effects of repetition and rule-familiarity. Psychol. Music **14**, 17–32

Smoliar, S.W. (1980): A computer aid for Schenkerian analysis. Comput. Music J. **4** (2), 41–48

Smoorenburg, G. (1970): Pitch perception of two-frequency stimuli. J. Acoust. Soc. Am. **48**, 924–942

Smoorenburg, G. (1972): Audibility region of combination tones. J. Acoust. Soc. Am. **52**, 603–614

Stevens, S.S. (1935): The relation of pitch to intensity. J. Acoust. Soc. Am. **6**, 150

Stevens, S.S. (1957): On the psychophysical law. Psychol. Rev. **64**, 153–181

Stevens, S.S., Volkmann, J. (1940): The relation of pitch to frequency. Am. J. Psychol. **53**, 329–353

Stevens, S.S., Volkmann, J., Newmann, E.G. (1937): A scale for the measurement of psychological magnitude pitch. J. Acoust. Soc. Am. **8**, 185–190

Stoll, G. (1982): "Spectral-pitch pattern: A Concept Representing the Tonal Features of Sounds", in *Music, Mind and Brain: The Neuropsychology of Music*, ed. by M. Clynes (Plenum, New York)

Stoll, G. (1983): Ambiguity of pitch of complex tones. Fourth Workshop on Physical and Neuropsychological Foundations of Music, Ossiach Austria

Stoll, G. (1984): Pitch of vowels: Experimental and theoretical investigation of its dependence on vowel quality. Speech Commun. **3**, 137–150

Stoll, G. (1985): Pitch shifts of pure and complex tones induced by masking noise. J. Acoust. Soc. Am. **77**, 188–192

Stoll, G., Parncutt, R. (1987): Harmonic relationships in similarity judgments of nonsimultaneous complex tones. Acustica **63**, 111–119

Stumpf, C. (1898): Konsonanz und Dissonanz. Beitr. Akust. Musikwiss. **1**, 1–108 [Cited in Plomp and Levelt (1965)]

Stumpf, C. (1909): Differenztöne und Konsonanz. Beitr. Akust. Musikwiss. **4**, 90–104 [Cited in Cazden (1954)]

Sundberg, J. (1982): "In Tune or Not? – A Study of Fundamental Frequency in Music Practise", in *Tiefenstruktur der Musik*, ed. by C. Dahlhaus, M. Kraus (Technical University of Berlin, Berlin)

Sundberg, J., Lindqvist, J. (1973): Musical octaves and pitch. J. Acoust. Soc. Am. **54**, 922–929

Sundberg, J., Askenfelt, A., Frydén, L. (1983): Musical performance: A synthesis-by-rule approach. Comput. Music J. **7** (1), 37–43

Sutton, R.A., Williams, R.P. (1969): Residue pitches from two-tone complexes. J. Sound. Vib. **13**, 195–199

Tanner, R. (1981): Psycharithms as the mathematical nature of music. J. Musicol. Res. **3**, 293–334

Tartini, G. (1754): *Trattato di musica secondo la vera scienza dell'armonia* (Padova) [Cited in Cazden (1954)]

Terhardt, E. (1968a): Über die durch amplitudenmodulierte Sinustöne hervorgerufene Hörempfindung. Acustica **20**, 210–214

Terhardt, E. (1968b): Über akustische Rauhigkeit und Schwankungstärke. Acustica **20**, 215–224

Terhardt, E. (1968c): Über ein Äquivalenzgesetz für Intervalle akustischer Empfindungsgrößen. Biol. Cybern. **5**, 127–133

Terhardt, E. (1969): Frequency analysis and periodicity detection in the sensations of roughness and periodicity pitch. Symposium on Frequency Analysis and Periodicity Detection in Hearing, Driebergen, The Netherlands.

Terhardt, E. (1970): Oktavspreizung und Tonhöhenverschiebung bei Sinustönen. Acustica **22**, 345–351

Terhardt, E. (1971): Die Tonhöhe harmonischer Klänge und das Oktavintervall. Acustica **24**, 126–136

Terhardt, E. (1972): Zur Tonhöhenwahrnehmung von Klängen. Acustica 26, 173–199

Terhardt, E. (1974a): Pitch, consonance and harmony. J. Acoust. Soc. Am. **55**, 1061–1069

Terhardt, E. (1974b): On the perception of periodic sound fluctuations (roughness). Acustica **30**, 201–213

Terhardt, E. (1976): Ein psycholakustisch begründetes Konzept der musikalischen Konsonanz. Acustica **36**, 121–137

Terhardt, E. (1977): "The Two-Component Theory of Musical Consonance", in *Psychophysics and Physiology of Hearing*, ed. by E.F. Evans, J.P. Wilson (Academic, London)

Terhardt, E. (1978): Psychoacoustic evaluation of musical sounds. Percept. Psychophys. **23**, 483–492

Terhardt, E. (1979a): Calculating virtual pitch. Hearing Res. **1**, 155–182

Terhardt, E. (1979b): Conceptual aspects of musical tones. Humanities Assoc. Rev. **30**, 45–57

Terhardt, E. (1982): "Die psychoakustischen Grundlagen der musikalischen Akkordgrundtöne und deren algorithmische Bestimmung", in *Tiefenstruktur der Musik*, ed. by C. Dahlhaus, M. Kraus (Technical University of Berlin, Berlin)

Terhardt, E. (1983): Musikwahrnehmung und elementare Hörempfindungen (with English tranlation). Audiol. Acoust. **22**, 53–56, 86–96

Terhardt, E. (1985): Fourier transformation of time signals: Conceptual revision. Acustica **57**, 242–256

Terhardt, E. (1986): Gestalt principles and music perception. In *Auditory Processing of Complex Sounds*, ed. by W.A. Yost, C.S. Watson (Erlbaum, Hillsdale, NJ)

Terhardt, E., Fastl, H. (1971): Zum Einfluß von Störtönen und Störgeräuschen auf die Tonhöhe von Sinustönen. Acustica **25**, 53–61

Terhardt, E., Grubert, A. (1987): Factors affecting pitch judgments as a function of spectral composition. Percept. Psychophys. **42**, 511–514

Terhardt, E., Seewann, M. (1983): Aural key identification and its relationship to absolute pitch. Music Percept. **1**, 63–83

Terhardt, E., Seewann, M. (1984): Auditive und objektive Bestimmung der Schlagtonhöhe von historischen Kirchenglocken. Acustica **54**, 129–144

Terhardt, E., Ward, W.D. (1982): Recognition of musical key: Exploratory study. J. Acoust. Soc. Am. **72**, 26–33

Terhardt, E., Zick, M. (1975): Evaluation of the tempered tone scale in normal, stretched and contracted intonation. Acustica **32**, 268–274

Terhardt, E., Stoll, G., Seewann, M. (1982a): Pitch of complex tonal signals according to virtual pitch theory: Tests, examples and predictions. J. Acoust. Soc. Am. **71**, 671–678

Terhardt, E., Stoll, G., Seewann, M. (1982b): Algorithm for extraction of pitch and pitch salience from complex tonal signals. J. Acoust. Soc. Am. **71**, 679–688

Terhardt, E., Stoll, G., Schermbach, R., Parncutt, R. (1986): Tonhöhenmehrdeutigkeit, Tonverwandtschaft und Identifikation von Sukzessivintervallen. Acustica **61**, 57–66

Thaut, M.H. (1985): The use of auditory rhythm and rhythmic speech to aid temporal muscular control in children with gross motor dysfunction. J. Music Therapy **22**, 108–128

Thomson, W. (1983): Functional ambiguity in musical structures. Music Percept. **1**, 3–27

Thompson, W.F., Cuddy, L.L. (1986): Local and acoustic factors in Bach Chorales. 12th International Congress on Acoustics, Toronto

Thurlow, W.R., Elfner, L.F. (1959): Continuity effects with alternately sounding tones. J. Acoust. Soc. Am. **31**, 1337–1339

Thurlow, W.R., Erchul, W.P. (1977): Judged similarity in pitch of octave multiples. Percept. Psychophys. **22**, 177–182

Thurlow, W.R., Rawling, L.L. (1959): Discrimination of number of simultaneously sounding tones. J. Acoust. Soc. Am. **31**, 1332–1336

Tjernlund, P., Sundberg, J., Fransson, F. (1972): "Grundfrequenzmessungen an schwedischen Kernspaltflöten", in *Musikhistoriska museets skrifter 4*, ed. by E. Emsheimer (Stockmann, Stockholm)

Tonkova-Yampol'skaya, R.V. (1973): "Development of Speech Intonation in Infants During the First Two years of Life", in *Studies of Child Language Development*, ed. by C. Ferguson, D. Slobin (Holt, Rinehart and Winston, New York)

Trehub, S.E. (1987): Infants' perception of musical patterns. Percept. Psychophys. **41**, 635–641

Ueda, K., Ohgushi, K. (1987): Perceptual components of pitch: Spatial representation using a multidimensional scaling technique. J. Acoust. Soc. Am. **82**, 1193–1200

Uhr, L. (1963): "Pattern recognition" computers as models for form perception. Psychol. Bull. **60**, 40–73

Valentine, C.W. (1914): The method of comparison in experiments with musical intervals and the effects of practice on the appreciation of discords. Br. J. Psychol. **7**, 118—135

Verny, T. (1981). *The Secret Life of the Unborn Child* (Delta, New York)

Vince, M.A. (1980): The posthatching consequences of prehatching stimulation: Changes with the amount of prehatching and posthatching exposure. Behaviour **75**, 36–53 [Cited in Vince et al. (1982a)]

Vince, M.A., Armitage, S.E., Walser, E.S., Reader, M. (1982a): Postnatal consequences of prenatal sound stimulation in the sheep. Behaviour **81**, 128–139

Vince, M.A., Armitage, S.E., Baldwin, B.A., Toner, J., Moore, B.C.J. (1982b): The sound environment of the foetal sheep. Behavior **81**, 296–315

Vitz, P.C. (1964): Preferences for rates of information presented by sequences of tones. J. Exp. Psychol. **68**, 176–183

Vos, J. (1982): The perception of pure and mistuned musical fifths and major thirds: Thresholds for discrimination, beats and identification. Percept. Psychophys. **32**, 297–313

Walliser, K. (1969a): Über die Spreizung von empfundenen Intervallen gegenüber mathematisch harmonischen Intervallen bei Sinustönen. Frequenz **23**, 139–143

Walliser, K. (1969b): Zur Unterschiedsschwelle der Periodentonhöhe. Acustica **21**, 329–336

Walliser, K. (1969c): Über die Abhängigkeiten der Tonhöhenempfindung von Sinustönen vom Schallpegel, von überlagertem drosselndem Störschall und von der Darbietungsdauer. Acustica **21**, 211

Ward, W.D. (1954): Subjective musical pitch. J. Acoust. Soc. Am. **26**, 369–380

Ward, W.D. (1962): On the perception of the frequency ratio 55:32. J. Acoust. Soc. Am. **34**, 679

Ward, W.D. (1963): Absolute pitch. Sound **2**, 14–41

Ward, W.D. (1970): "Musical Perception", in *Foundations of Modern Auditory Theory,* ed. by J.V. Tobias (Academic, New York)

Ward, W.D., Burns, E.M. (1982): "Absolute Pitch", in *The Psychology of Music*, ed. by D. Deutsch (Academic, New York)

Ward, W.D., Martin, D.W. (1961): Subjective evaluation of musical scale temperament in pianos. J. Acoust. Soc. Am. **33**, 582–585

Watson, R.W. (1982): *Viennese Harmonic Theory from Albrechtsberger to Schenker and Schoenberg* (U.M.I. Research Press, Ann Arbor, MI)

Watt, H.J. (1917): *The Psychology of Sound* (Cambridge University Press, Cambridge) [Cited in Ward (1970)]

Webster, J.C., Schubert, E.D. (1954): Pitch shifts accompanying certain auditory threshold shifts. J. Acoust. Soc. Am. **26**, 754–758 [Cited in Ward (1970)]

Webster, J.C., Miller, P.H., Thomson, P.O., Davenport, E.W. (1952): The masking and pitch shifts of pure tones near abrupt changes in the thermal noise spectrum. J. Acoust. Soc. Am. **24**, 147–152

Wedin, L., Goude, B. (1972): Dimensional analysis of the perception of timbre. Scand. J. Psychol. **13**, 228–240

Wegel, R.L., Lane, C.E. (1924): The auditory masking of one sound by another and its probable relation to the dynamics of the inner ear. Phys. Rev. **23**, 266–285

Wertheimer, M. (1923): "Principles of Perceptual Organization", in *Readings in Perception*, ed. by D.C. Beardslee, M. Wertheimer (Van Nostrand, Princeton, NJ 1958)

Wessel, D. (1979): Timbre space as a musical control structure. Comput. Music J. **3**, 45–52 [Cited in Singh (1987)]

West, R., Cross, I., Howell, P. (1987): Modelling music as input-output and as process. Psychol. Music Music Educ. **15**, 7–29

Wever, E.G., Bray, C.W. (1937): The perception of low tones and the resonance-volley theory. J. Psychol. **3**, 101–104

Whitfield, I.C. (1967): *The Auditory Pathway* (Arnold, London) [Cited in Moore (1982)]

Wickelgren, W.A. (1966): Consolidation and retroactive interference in short-term recognition memory for pitch. J. Exp. Psychol. **72**, 250–259

Wickelgren, W.A. (1969): Associative strength theory of recognition memory for pitch. J. Math. Psychol. **6**, 13–61

Wienphal, R.W. (1959): Zarlino, the *sensario*, and tonality. J. Am. Musicol. Soc. **12**, 27–41

Wightman, F.L. (1973): The pattern-transformation model of pitch. J. Acoust. Soc. Am. **54**, 407–416

Wilding-White, R. (1961): Tonality and scale theory. J. Music Theo. **5**, 275–286

Williams, P. (1968): Equal temperament and the English organ 1675–1825. Acta Musicol. **40**, 53–65

Wood, A. (1961): *Physics of Music* (Dover, New York)

Wright, J.K. (1986): "Auditory Object Perception: Counterpoint in a New Context"; Masters dissertation, McGill University, Montreal

Yasser, J. (1929): *A Theory of Evolving Tonality* (American Library of Musicology, New York)

Young, R.W. (1939): Terminology for logarithmic frequency units. J. Acoust. Soc. Am. **11**, 134–139

Yunik, M., Swift, G.W. (1980): Tempered music scales for sound synthesis. Comput. Music J. **4** (4), 60–65

Zenatti, A. (1985): The role of perceptual-discrimination ability in tests of memory for melody, harmony and rhythm. Music Percept. **2**, 397–404

Zwicker, E. (1960): Ein Verfahren zur Berechnung der Lautstärke. Acustica **10**, 304–308

Zwicker, E. (1961): Subdivision of the audible frequency range into critical bands (Frequenzgruppen). J. Acoust. Soc. Am. **33**, 248

Zwicker, E. (1970): Masking and psychological excitation as consequences of the ear's frequency analysis, in *Frequency Analysis and Periodicity Detection in Hearing*, ed. by R. Plomp, G.F. Smoorenburg (Sitjthoff, Leiden) [(Cited in Moore (1982)]

Zwicker, E. (1962): *Psychoakustik* (Springer, Berlin, Heidelberg)

Zwicker, E., Feldtkeller, R. (1967): *Das Ohr als Nachrichtenempfänger*, 2nd ed. (Hirzel, Stuttgart)

Zwicker, E., Herla, S. (1975): Über die Addition von Verdeckungseffekten. Acustica **34**, 89–97

Zwicker, E., Scharf, B. (1965): A model of loudness summation. Psychol. Rev. **72**, 3–26

Zwicker, E., Terhardt, E. (1980): Analytical expressions for critical-band rate and critical bandwidth as a function of frequency. J. Acoust. Soc. Am. **68**, 1523–1525

Zwicker, E., Flottorp, G., Stevens, S.S. (1957): Critical bandwidth in loudness summation. J. Acoust. Soc. Am. **29**, 548–557

Subject Index

Entries also listed in the Glossary are marked "G".